KV-475-357

DESIGN, CONSTRUCTION AND REFURBISHMENT OF LABORATORIES

ELLIS HORWOOD SERIES IN INFORMATION SCIENCE

COMMUNICATION, STORAGE AND RETRIEVAL OF CHEMICAL INFORMATION
Editors: J. E. ASH, P. A. CHUBB, S. E. WARD, S. M. WELFORD, P. WILLETT

CHEMICAL INFORMATION SYSTEMS
J. ASH, Information Services Consultant and E. HYDE, ICI Pharmaceuticals Division

CHEMICAL NOMENCLATURE USAGE
R. LEES and A. SMITH, Laboratory of the Government Chemist, London

DESIGN, CONSTRUCTION AND REFURBISHMENT OF LABORATORIES
Editors: R. LEES and A. SMITH, Laboratory of the Government Chemist, London

70 0181088 9   TELEPEN

WITHDRAWN
FROM
UNIVERSITY OF PLYMOUTH

Charles Seale-Hayne Library
University of Plymouth
(01752) 588 588
LibraryandITenquiries@plymouth.ac.uk

DESIGN, CONSTRUCTION AND REFURBISHMENT OF LABORATORIES

Editors:

R. LEES, C.Chem., M.R.S.C., F.R.S.H., A.I.F.S.T.
Superintendent of Technical Administration

and

A. F. SMITH, C.Chem., M.R.S.C.
Head of Training, Information and Publicity

both of The Laboratory of the Government Chemist, London

Published by
ELLIS HORWOOD LIMITED
Publishers · Chichester

For the
LABORATORY OF THE
GOVERNMENT CHEMIST
London

First published in 1984 by
ELLIS HORWOOD LIMITED
Market Cross House, Cooper Street, Chichester, West Sussex, PO19 1EB, England

The publisher's colophon is reproduced from James Gillison's drawing of the ancient Market Cross, Chichester.

Distributors:

Australia, New Zealand, South-east Asia:
Jacaranda-Wiley Ltd., Jacaranda Press,
JOHN WILEY & SONS INC.,
G.P.O. Box 859, Brisbane, Queensland 40001, Australia

Canada:
JOHN WILEY & SONS CANADA LIMITED
22 Worcester Road, Rexdale, Ontario, Canada.

Europe, Africa:
JOHN WILEY & SONS LIMITED
Baffins Lane, Chichester, West Sussex, England.

North and South America and the rest of the world:
Halsted Press: a division of
JOHN WILEY & SONS
605 Third Avenue, New York, N.Y. 10016, U.S.A.

© 1984 Ellis Horwood Limited/Crown Copyright

British Library Cataloguing in Publication Data
Design, construction and refurbishment of laboratories. —
(Ellis Horwood series in information science)
1. Chemical laboratories
I. Lees, Ron II. Smith, A.F. (Arthur Francis)
III. Laboratory of the Government Chemist
542'.1 QD51

Library of Congress Card No. 84-15788

ISBN 0-85312-645-3 (Ellis Horwood Limited)
ISBN 0-470-20133-9 (Halsted Press)

Typeset by Ellis Horwood Limited.
Printed in Great Britain by R.J. Acford, Chichester.

COPYRIGHT NOTICE —
All Rights Reserved. No part of this publication may be reproduced, stored in a retrieval system, or transmitted, in any form or by any means, electronic, mechanical, photocopying, recording or otherwise, without the permission of Ellis Horwood Limited, Market Cross House, Cooper Street, Chichester, West Sussex, England.

Table of Contents

PLYMOUTH POLYTECHNIC
LIBRARY

Accn.
No.   · 181088 -9

Class
No.   542. 1   DES

Contl.
No.   0853126453

8 Contents

Foreword

A majority of workers in laboratories suffer from ill-designed accommodation inherited from their predecessors. Most scientists, whatever their discipline or product area, will be concerned with the design of new laboratories or major refurbishment perhaps on only one occasion during their career. The cost of providing these facilities is extremely high and organisations, whether in Government or private companies, have the right to expect that the money is spent wisely and effectively. Additionally, the design of new facilities must meet the evolving programme of the establishment in an efficient and cost-effective manner over a period of many years. Valuable staff time and finance is being wasted throughout the world because the expertise available in the design and construction of laboratories is not being shared. The editors and contributors have produced a book which will contain relevant information for all types of readers, be they educationalists or researchers, or involved in quality control, and whether they are located in developed or developing countries.

I consider it important to encourage the exchange of information across national boundaries on matters concerning scientific management and administration, and this book will, I believe, contribute to that dialogue.

Both editors are on the staff of the Laboratory of the Government Chemist and the work has been undertaken as part of the Laboratory's programme of technology transfer. The book should have wide appeal to the scientific and technical community. Authors include working scientists and educationalists who have contributed from their experiences in laboratory design, and specialists from architectural and engineering design practices responsible for implementing the requirements of laboratory staff and management.

Dr Ronald Coleman
Government Chemist
London

August 1984

The Contributors

Mr J. H. Armstrong started work with Duff and Geddes, Consulting Engineers, on hydroelectric projects and general structural engineering before joining Rendel, Palmer and Tritton to work on heavy foundation matters concerned with major overseas projects. His later career was with Associated Portland Cement Manufacturers, Harris and Sutherland and John Mowlens.

Mr Armstrong is now the Civil and Structural Engineering Partner with the Building Design Partnership (BDP) and has many years experience in the direction and management of major projects in the United Kingdom. He was project manager leading the team which developed the plans for the Channel Tunnel English terminal and is currently responsible for commissions in the City of London and at Colindale.

Mr P. Boardman is a Professional and Technology Officer with the Crown Suppliers (formerly Property Services Agency (PSA) Supplies) where he is concerned in a practical and advisory capacity with technical detail related to the design, development and installation of fume cupboards. He graduated in engineering and in his earlier career gained experience, mainly in development work, in the photolitho and aircraft industries.

Dr P. R. Boyce graduated at the University of Reading where he later gained a Ph.D. He joined the Electricity Council Research Centre, Capenhurst, as a Research Officer working on human responses to lighting conditions. His major areas of investigation have included the effects of lighting conditions on visual fatigue, the influence of age on visual performance, the effect of light source colour properties on hue discrimination, and the variability and utility of contrast rendering factor. He has also been involved in thermal comfort research, surveys of occupants' attitudes to deep plan office buildings and more recently, assessments of noise disturbance caused by domestic heat pumps. Dr Boyce has contributed more than forty papers to scientific journals and conferences, and written a book entitled *Human Factors in Lighting*.

Dr Boyce has been a member of the Technical Committee of the Illuminating Engineering Society (now the Chartered Institution of Building Services Lighting Division) since 1969 and is currently chairman of the panel which has produced the Chartered Institution of Building Services' (CIBS) *Code for Interior Lighting*, 1984.

Mr. A. J. Branton joined the Laboratories Investigation Unit of the Department of Education and Science (DES) in 1968 following periods in private practice and the University Grants Committee. He played a key role in the early research and systems design programmes of the Unit, and was site architect for the Unit's major development project, the Charles Darwin Building, Bristol Polytechnic.

Mr Branton has been principal architect with the Unit since 1975 working on projects and systems for the reorganisation and conversion of existing buildings for laboratory use. He has advised educational and industrial organisations in the UK and overseas on the planning of science buildings and the design of adaptable laboratories. He is author of a number of papers on laboratory design and is currently writing a book on this topic.

Mr R. E. Clynes qualified as an architect at the Regent Street Polytechnic following which he worked for a number of architects on projects ranging from London Airport to various educational buildings, and was site architect for a hospital in the Arabian Gulf. In 1964 he became principal architect for the University Grants Committee, having a particular responsibility for development work.

Since 1968 Mr Clynes has been in the Architects and Building Group at DES where he is responsible for all development work in the area of higher and further education. Since 1978 he has been also Director of the Laboratories Investigation Unit.

Professor Miles Danby is Head of Department and Professor of Architecture at the University of Newcastle upon Tyne. He combines this post with the directorships of the Project Office and the postgraduate courses in Housing for Developing Countries. From 1965 to 1970 he was Professor of Architecture and Head of Department at the University of Khartoum. Before taking up his post at Khartoum he had been on the staff and subsequently Head of Department at the University of Science and Technology at Kumasi, Ghana.

Professor Danby has been involved in the practice of architecture as well as teaching and has been responsible for buildings and projects in Ghana, Saudi Arabia, Spain, Sudan and the United Kingdom. He has also been consultant to many organisations including the British Council, ODA, UNESCO, FAO and overseas governments.

Professor Danby has written papers, articles and a book on the architecture and environment of the Third World and is currently associate editor of the *Third World Planning Review*.

Mr P. E. Donnachie is a Principal Mechanical and Electrical Engineer with PSA. He has had many years experience in the design, construction, testing and inspection of electrical installations in a practical and advisory capacity. Mr Donnachie has served on a number of committees concerned with electrical standards.

Mr H. B. Ellwood joined BDP directly from the University of Manchester School of Architecture. He was at the Rome office of BDP for two years during which he worked on projects in Belgium, Italy and Switzerland. Mr Ellwood has worked on housing, schools, churches, hospitals, administrative and industrial buildings and university residences. He is currently working on microbiological laboratories, a production centre and a pilot fermentation plant at Porton Down in Wiltshire.

Mr M. S. Griffiths worked in various laboratories including microbiology laboratories on infectious diseases before training as a Specialist Inspector with the Health and Safety Executive (HSE). He specialises in advice and enforcement in premises where infectious hazards occur. This work covers hospital, university and industrial laboratories and includes very dangerous pathogen laboratories and wards where Lassa Fever and rabies samples might be handled. Mr Griffiths is now with the Occupational Health and Safety Service of the Scottish and Newcastle Breweries plc, Edinburgh.

Mr L. G. Haley is commercial manager at Cygnet Joinery Limited a company with which he has been associated since 1938, though during this period he spent about ten years at Shapland and Petter, a company manufacturing high quality furniture largely for municipal and private buildings, where he established laboratory furniture production. He has been involved in handling numerous projects for major industrial and institutional customers and his experience includes investigation and market survey work in the Middle East particularly the Arab Emirates and the Lebanon.

Mr N. H. E. Hanby graduated in Civil Engineering at Bristol University and Lecturer in Physical Chemistry specialising Analytical and Radiochemistry at Birkenhead Technical College and then joined the Department of Zoology at the University of Sheffield. In 1977 he was appointed Safety Director at Imperial College, London.

Mr Hanby has served on the Technical Committee of the National Water Council Fittings Scheme and the Standing Technical Committee on Water Regulations, which advises the Department of the Environment.

Dr G. Hargreaves graduated from Liverpool University in 1955 to join the United Kingdom Atomic Energy Authority (UKAEA). He later became Senior Lecturer in Physical Chemistry specialising in analytical and radiochemistry at

Birkenhead Technical College and then joined the Department of Zoology at the University of Sheffield. In 1977 he was appointed Safety Director at Imperial College, London.

Mr A. G. Harris is Head of Technical Services and Planning at the Laboratory of the Government Chemist (LGC) which he joined in 1979. He has been concerned with the design of chemistry and physics laboratories since 1956 and has acted as client's officer in several new building schemes in the Government Service.

Mr Harris has been engaged in the planning and design of a new laboratory for LGC and also with design consultancy on chemistry laboratories for other organisations. He has wide experience also in cryogenics and generation of high magnetic fields.

Mr T. Henney trained as an Architect at the School of Architecture at Edinburgh College of Art and at the Graduate School of Architecture, Massachusetts Institute of Technology, USA. He was formerly Principal Architect with the Architecture Research Unit, University of Edinburgh, but is now a principal in private practice and a member of the Quoin Architects Group, Edinburgh. Mr Henney has been responsible for the design of research laboratories for the University of Edinburgh, the Medical Research Council and the Property Services Agency. He has written a number of articles on the subject of laboratory design.

Mr K. C. Hignett graduated in electrical power engineering in 1945 and subsequently gained wide experience in the chemical and oil industries as a control and instrumentation engineer. He entered UKAEA in 1956 and in 1973 joined the Systems Reliability Service of the Authority at Culcheth in Cheshire where he is at present a senior reliability engineer. Mr Hignett's current professional activities cover a wide field of chemical materials processing and handling, including water treatment in public supplies and waste treatment. He has contributed papers for a number of conferences and seminars.

Mr W. R. Hodgkins is Head of the Theoretical Section at the Electricity Council Research Centre, Capenhurst, being responsible for mathematics and computing services. After graduating in mathematics at Cambridge University he joined the Industrial Group of UKAEA working on the behaviour of fuel elements and carrying out some of the first deformation analyses performed by computer. During this period he was seconded for a year as a Research Associate with the Division of Engineering at Brown University. In 1967 he took up the post of Head of the Numerical Analysis Section at the Nelson Research Laboratories of English Electric Company before joining the Electricity Council in 1969.

Mr W. P. Horsnell is a Chartered Quantity Surveyor who worked in the Ministry of Housing and Local Government before moving to the Department of the

Environment. He continued his career in the University Grants Committee from 1971 to 1981 when he joined DES to work in the Laboratories Investigation Unit.

Mr P. J. Jackman spent three years as a project engineer concerned with aircraft oxygen and air-conditioning systems at Normalair Ltd, after graduating in Mechanical Engineering at Brunel University. In 1966 he was appointed as a Senior Research Officer at the then Heating and Ventilating Research Association, now the Building Services Research and Information Association (BSRIA), and has specialised in the study of the movement of air into and within buildings. His work is concerned with the provision of conditions acceptable to occupants or processes in buildings of various types, and with the energy-related consequences of ventilation and air infiltration. Mr Jackman's involvement in this area continues with project management responsibilities as Deputy Head of the Technical Division of BSRIA.

Mr R. A. J. Kinsey is Managing Director of Bluemain Ltd, a services company specialising in the specification, design and construction of laboratories. He was for many years an engineering manager with British Oxygen Co. Ltd concerned with the design of both research and production laboratories and general gases engineeering. His present interests cover energy conservation in particular and consulting engineering in general.

Professor E. R. Laithwaite was born in Atherton, Lancashire, and educated at Kirkham Grammar School and the Regent Street Polytechnic (now the Polytechnic of Central London). He joined the RAF in 1941 and from 1944 to 1945 worked on automatic pilots at the Royal Aircraft Establishment, Farnborough. After the war he graduated at Manchester University, subsequently being appointed Lecturer, then Senior Lecturer, in Electrical Engineering. In 1964 he became Professor of Heavy Electrical Engineering at Imperial College, London, and also held the post of External Professor of Applied Electricity at the Royal Institution from 1967 to 1976. A Director of Linear Motors Ltd, Landspeed Ltd and Cotswold Research Ltd, he also consults for a large number of industrial organisations on an ad hoc basis.

In 1966 he was awarded the Royal Society's S. G. Brown Gold Medal for inventions relating to linear induction motors. He has written over two hundred and fifty papers and articles and ten books on electrical engineering and allied subjects.

Mr A. G. Lewis worked in the steel industry, holding a variety of posts in the three Welsh tinplate works, before joining Amersham International plc (then The Radiochemical Centre Ltd) in 1977 as Engineering Manager. He was initially responsible for setting up the engineering facilities at the new Cardiff site, built between 1976 and 1980, and was actively involved in all aspects of the construction and commissioning of the laboratory buildings and their associated services.

His present responsibility is for all site engineering activities, as well as supplies, radiological safety and effluent disposal.

Dr A. L. Longworth is a consultant ventilation engineer with special interest in the control of air flows in enclosed spaces. Throughout his career as a Chartered Engineer he has been concerned in the design, installation, and commissioning of building services engineering projects, interposed with periods of service as an engineer in the then Ministry of Fuel and Power, and as a Senior Lecturer in building services engineering in the University of Manchester.

His recent research activities have included aerodynamics of buildings, temperature and dust level control in animal rooms, and aerodynamic design and performance testing of laboratory fume cupboards.

Dr Longworth is a member of the British Standards Institution (BSI) committee responsible for fume cupboards, and chairman of a joint CIBS and HSE committee on the control of airborne contaminants.

Dr J. R. Moody was born and raised in Richmond, Virginia, USA, graduated in chemistry at the University of Richmond and later gained a Ph.D. at the University of Maryland. He worked in industry and then served as a post-doctoral lecturer before joining the National Bureau of Standards (NBS) as a research chemist in 1971. Dr Moody's research interests have centred on trace analysis of metals and he has designed or built a number of laboratories designed specifically for trace analysis work. For the last seven years, Dr Moody has been the Safety Officer for the NBS Centre for Analytical Chemistry and in that capacity has served as a liaison between the architectural engineering scientists and safety and health officers at the establishment.

Dr Moody is a member of the American Chemical Society and the Society for Applied Spectroscopy and has been President of the Chemical Society of Washington.

Mr D. L. Morgan began his career in drug research and development and later moved to management services at Beecham Research Laboratories. In 1976 he joined Huntingdon Research Centre where he is currently Divisional Commercial Manager. While at Huntingdon he has been responsible for the research and design of several laboratory installations ranging from a specialised mutagenicity unit for testing carcinogenic materials to an extensive laboratory block for biochemistry and radiochemical work.

Mr N. H. Pearce has been in charge of the Safety Section, which he started, of the University of Bristol for the last ten years. He is a past Chairman of the Universities Safety Association and present Chairman of the Institute of University Safety Officers. He is a member of the Health, Safety and Environment Committee of the Royal Society of Chemistry, the Committee of Vice-Chancellors' and Principals' University Committee on Safety and the BSI Committee concerned with laboratory fittings. Mr Pearce was a Regional/

District member of Council of the Royal Institute of Chemistry from 1969 to 1972.

Dr M. E. Peel is the Safety Adviser to Glaxo Group Research at Ware. Previously he was Administrator of their Chemistry Division and a Research Leader in chemistry in the Research Division of Allen and Hanburys Ltd.

From 1972 to 1974, Dr Peel was Chairman of the Fine Chemicals Group of the Society of Chemical Industry.

Mr R. C. Rawling joined the Post Office (now British Telecom) in 1942 and has had wide interests ranging from the design of underground cable networks to electronic letter sorting machinery. He currently heads the group at British Telecom headquarters responsible for cabling work in customers' premises. This includes the traditional analogue services, cable TV, digital services at all speeds and many of the practical interfaces needed as a result of liberalisation of the British Telecom monopoly.

Dr F. D. Riley is Head of Planning and Management Services at the Electricity Council Research Centre, Capenhurst, being responsible for research planning, information services, and various administrative functions. After graduating in chemistry from the University College of North Wales, he spent two years as a research chemist with Pilkington Brothers R&D Department. He then went up to Oxford University to carry out postgraduate research in high temperature electrochemistry. In 1969, he returned to Pilkingtons, initially as a Senior Technologist, then as a Manager in R&D Planning. In 1977 he joined the Vickers Group as Budgets, Information and Planning Manager with the Roneo-Vickers Division and in 1979 he took up his present post with the Electricity Council.

Dr A. G. Robins studied Aeronautics at Imperial College, London, where he also gained a Ph.D. for studies of turbulence in jet flows. He joined the Central Electricity Generating Board (CEGB) in 1969 to work in the wind tunnel group at the Marchwood Engineering Laboratories, and in 1983 moved to the Central Electricity Research Laboratories as group leader in the new stratified flow laboratory. Whilst with the CEGB, Dr Robins's research has been concerned mainly with laboratory-scale simulation, and theoretical modelling of atmospheric flow and dispersion problems.

Ms Beverley Sayers graduated in physics from Lancaster University. After some years working at Warrington Research Centre carrying out research into fire and associated hazards she moved to the consultancy field where she became involved in the safety and reliability analysis of major industrial plants for the process industries. As an independent consultant to the Systems Reliability Service of the UKAEA she worked on projects for the chemical industry, off-shore oil installations, gas terminals and major government departments. She is at present employed by the National Nuclear Corporation Departments working on the safety and reliability of protection systems designed for the UK Advanced Gas Cooled Reactors. As a member of the Safety and Reliability Society she is involved with the general organisation and in particular for the activities of the North West Branch of the Society.

Dr R. C. Slade graduated in chemistry at the University of Hull and has worked in the research laboratories at the Universities of Essex and Manchester, and at University College London. In 1969 he was appointed a lecturer in chemistry at Queen Elizabeth College, London. His interest in safety began in 1972 when he became Safety Officer in the Chemistry Department and for six years he chaired the college's Safety Committee. In 1978 he was appointed College Safety Officer but still continues to lecture in chemistry.

As College Safety Officer he has set up and continues to organise the college's arrangments for dealing with waste chemicals and solvents. Dr Slade is a member of the University of London's Standing Committee of Safety Officers and serves on a working party looking at design and performance of fume cupboards. Amongst his other interests is the application of analytical techniques to the study of laboratory atmosphere pollution.

Mr K. Oldham Smith has had a lifetime's experience of electrical installation practice in industrial, commercial and government establishments. Since 1967, when he became an HM Senior Electrical Inspector of Factories, he has specialised in electrical safety on which he lectures and writes. Until his retirement in 1980, he represented the HSE on a number of BSI electrical committees. He is now in private practice as an electrical safety consulting engineer.

Dr H. Spencer graduated from Liverpool University and went to Canada to undertake post-doctoral research and later teach chemistry at the University of Saskatchewan.

On his return to the United Kingdom in 1969, Dr Spencer joined ICI Pharmaceutical Laboratories and later became the Scientific Safety Adviser. In 1980 he became Senior Safety Officer at the Frythe site of Smith, Kline and French Research Ltd where he is responsible for co-ordinating general safety and emergency services and providing technical safety services.

Mr W. A. Stevenson was trained in Liverpool in construction management. He worked with Bovis and later Tysons (Contractors) Ltd, with whom he was involved in a number of complex construction projects. Mr Stevenson joined Leyland Vehicles in 1978 as a 'clients project manager', responsible for the construction of a major engineering test centre. He was a project manager with Wimpey Construction (UK) Ltd and is now Managing Director of Bellway Urban Renewal (Northern) Ltd.

Mr R. Taylor graduated from Sheffield University after working for the National Coal Board. He joined UKAEA in 1955 and has worked on the electrochemistry of plutonium in fused salt systems. His work on reactor safety studies includes fission product release from overheated thermal reactor fuels, development of charcoals for trapping iodine-131 and special filters for liquid sodium metal aerosol trapping. Since 1970 Mr Taylor has been concerned with fast reactor fuel studies including the development of new routes for fabrication of (UPu)C fuels and basic research on $(UPu)O_2$ fuels.

Mr. L. Thomas, a chartered mechanical engineer, joined Shell Research Ltd in 1949 after service in the RAF and two years with a local authority. Until 1964 he was engaged in the design and construction of large special-purpose research rigs. Subsequently he has been concerned with setting up a major research centre from a modest base. This work involved the design, construction and operation of research facilities concerned mainly with chemistry, biology and toxicology, plus ancillary features which support a large scientific complex. Mr Thomas retired in 1983 but retains interest in laboratories as a part-time consultant.

Dr W. R. Tully joined Roussel Laboratories Ltd in 1970 after studying chemistry at Oxford University. In 1977 he helped to design and set up a large-scale laboratory for the Company's Chemistry Department. Currently, Dr Tully is a Project Leader for chemical research.

Mr J. Weeks was Deputy Director of the Nuffield Foundation's Division for Architectural Studies from 1950 to 1960. This team carried out basic research on hospital design and undertook studies on the design of research laboratories. His work with the team identified problems of growth and change now central to architecture.

In 1960 the firm of Llewelyn-Davies Weeks was formed and Mr Weeks has been responsible for most of the firm's work in hospital and laboratory building. He designed Britain's national medical research complex – an integrated medical care and research centre at Northwick Park – and a wide range of medical research buildings in many parts of the world.

Mr Weeks is the author of many publications and has lectured extensively in the United Kingdom and in other countries on subjects connected with the design of hospitals and laboratories.

Mr B. G. Whitehouse is Chief Quantity Surveyor at DES where he has worked since 1962. He has responsibility for providing building economic services to that Department's Laboratory Investigations Unit in its work on the design of laboratories.

Mr R. F. Young joined Roussel Laboratories Ltd in 1968 after six years in the dairy industry. He has held various engineering and building posts in the Company and is now Facilities Planning Manager, responsible for capital projects and fabric maintenance.

The opinions expressed by the authors of the chapters in this book are their own and do not necessarily represent those of their organisations.

Introduction

Changing requirements for laboratory accommodation

A laboratory, as defined in the dictionary, is a building set apart for experiments in natural science. Originally this applied to alchemy and subsequently chemistry, but the present concept covers all disciplines — chemistry, biochemistry, medicine, physics, engineering, etc. The term may be applied to any area where experimental work is carried out and covers any facility from a single room to a large building complex or Science Park.

The design and layout of laboratory accommodation has evolved continuously throughout the centuries in an endeavour to provide facilities which are suitable for the purpose required, safe to work in and economical to construct and use. This process of change has accelerated in recent years as new methods for scientific investigation and control have become available. With these new techniques there is generally much less emphasis on manipulative 'bench skills' and more on instrumental methods requiring a supervisory scientist who controls the experiments and interprets the results. Engineering research often requires large-scale facilities which must, for example, incorporate the ability to simulate adverse weather or operating conditions. Health and radiochemical laboratories must comply with stringent safety procedures which are designed to protect both the worker and the integrity of the sample under test. The trend in analytical chemistry has been away from 'wet' chemistry towards instrumentation and only the preparative stage requires such traditional methods as acid digestion or solvent extraction. They physicist demands both an improved and a wider range of facilities in which to undertake research or to control the quality of manufacturing processes which involve the new technologies of electronics and microprocessor control.

The provision of laboratory accommodation is governed by the cost of providing these facilities and by their rapid obsolescence. Until the early part of the twentieth century laboratory requirements could be specified with the reasonable certainty that the accommodation would meet the needs of staff and programmes for at least forty or fifty years.

The rapid change in scientific methods and knowledge has altered this situation. New apparatus places increasing demands on space and often requires some form of environmental control. The ratio for allocating space between benchwork and free-standing items has been subject to considerable variation. Research workers now demand more space to be allocated for their personal needs as new equipment is designed and becomes dedicated to a single project. The implications on accommodation are far-reaching. Even minor changes in the configuration of rooms can, with the considerable increase in building costs in recent years, add significantly to overheads.

To overcome this problem the designer has introduced flexibility into the plans with such features as the use of moveable benches, and partition units with take-off points for a wide range of services. This concept is not wholly acceptable for all facets of scientific work since safety requirements must always be paramount. Most design solutions for laboratories will be a compromise between the requirements of the scientist, safety, flexibility, finance and the availability of space on the site or in existing accommodation for conversion. Each action taken is likely to have implications on other aspects of the design and construction. An increase in the provision of safe fume enclosures for work involving toxic materials will have an adverse effect on any energy conservation measures and thus on the total operating costs of the laboratory. Increased requirements for natural light will alter the level of solar gain and possibly result in significant changes in any plans for control of the laboratory environment.

Safety requirements have become more stringent and impose their own constraints on the design of the accommodation. The use of gas cylinders within the laboratory area may be considered inadvisable by government officials charged with ensuring the safety of the staff. Writing areas may have to be separated from working areas and housed elsewhere within the laboratory if there is a risk of accident. Blast partitioning will be needed for certain types of operation. The trend in design has therefore been away from small rooms with space for one or two scientists or technologists towards large, open-area laboratories where the safety of individual workers can be monitored by other staff using the accommodation.

The cost of providing new or even refurbished laboratories can be a major inhibitory factor for many companies and organisations. Innovatory methods have therefore been introduced in the design process, through repetition in layout, and in the manner of construction by using off-site construction methods. The possibilities for a revolution in techniques in the provision of new laboratory space are not as great compared with, for example, the revolutionary robotic factory being introduced within manufacturing industry. Construction work on laboratories still relies heavily on human labour and often demands the use of materials which must be manufactured or fabricated to high standards. Design needs are intricate when compared with those required for a warehouse or simple office and detailed discussion must take place between architects, design engineers and clients. A further factor to be considered is that, once an organisation has decided to commission a new laboratory, it can become a show place. It is an area where the Directorate team can show visitors that they have actively considered the future prosperity of the company or country through their research, or demonstrate their concern for quality and product development.

Potential users of both refurbished and new accommodation can find themselves faced with competing interests. They require accommodation which will meet their immediate short-term requirements and which can be smoothly and efficiently converted to take account of changes in both scientific techniques and programmes. The accommodation must be a pleasant and stimulating environment in which to spend the working day. Against these needs are those of the financial controller who will necessarily have to set a limit on the cost of construction, and the architect who is concerned with the overall appearance of the building and the harmony of the design with its surroundings. The builder will look for a design which keeps down construction costs and minimises on-site problems. Consideration will have to be given to suppliers who need to deliver materials to the site during construction and later to the completed building, and to the service engineer who will have to maintain the, often elaborate, environmental and other control schemes introduced into the design.

A new laboratory block or complex contains many features not directly connected with experimental work and the design team will require guidance on a wide range of needs, examples of which are indicated in Table I.1; the list is not comprehensive. As a result the client's representatives in any discussions find themselves faced with a multiplicity of decisions as to what shall or shall not be included. Their experience in a particular scientific discipline may have provided little help on this and, if the final product to be successful and acceptable, thorough discussion with a free interchange of ideas between all parties will be necessary.

Table I.1

Ancillary requirements in a laboratory complex

Archive storage

Canteen
Car parks
Cleaner's stores and mess area
Cloakrooms
Committee and conference rooms
Computer demonstation/training area
Computer rooms
Conference facilities (includes projection room(s), storage area, etc.)

Equipment stores
Exhibition area

Table I.1 – *continued*

File storage
First-aid rooms

Garaging and transport repair facilities
Gas cylinder store and distribution system
Goods delivery and reception

Interview rooms

Library
Library store
Lifts
Loading bays

Main reception area
Maintenance engineers and ancillary control areas
Mess areas
Messengers' room

Offices

Post room
Printing   rographic rooms

Redi   ipment store
Refi   tion

Security control area
Secure storage
Staff and/or trades union office
Stores (chemical, solvent, special, furniture, toxic and hazardous materials,
    waste, etc.)

Telephone exchange and operators' room
Toilets
Training rooms/laboratories

Waste disposal
Workshops

The design team will also require information on such features as staff, traffic and visitor movement, the usage of special facilities such as the canteen and library, and any restrictions to be placed on access to areas within the complex. Additional features and services that may be required at a future date must also be discussed at the design stage.

The client must consider all options before meeting the design team and have a clear view of the type of accommodation that is required. A primary consideration is whether the accommodation could be achieved by the more effective use of existing space given that some building adaptation could take place. Disruption costs should be included with building costs when this approach is chosen. The second consideration is whether the proposed accommodation could be achieved only by the construction of a new building, or whether an existing building or site could be modified for the purpose. The cost of modifying existing buildings can be high and may approach those for a new building.

There are further options available. The new building can be leased or purchased. It is possible that a developer, or a local authority or body in a less-favoured area or new town could be persuaded to erect a building to an agreed specification and then rent or sell. Another possibility is that the building could be erected on land owned or purchased for other purposes by other organisations within the client's company.

It is probable that it will be the first time that some or all members of the team will have had to design a laboratory. In this case early misconceptions which relate design features to those commonly used in a modern office block or large warehouse will have to be dispelled. Such factors as the need for a vibration-free structure, the effect of the building frame on electronic instruments, the need for north light where required, the corrosive effect of fumes, etc. will all need to be explained, and controlled throughout the project.

Most projects follow a conventional development pattern as indicated in Table I.2.

The time taken to convert an existing area or provide a new laboratory is often far longer than expected. A major refurbishment or conversion of an existing building can take from one to three years. For a new building, from a point at which a decision is taken to appoint architects to the final occupation, the time span may be from two-and-a-half to five years, depending on such factors as the complexity of the services, the need to purchase land, difficulties experienced in obtaining planning permission, and effluent disposal problems. Even leasing an existing building may take more than a year to achieve.

It was against this background that the Laboratory of the Government Chemist in association with the Chartered Institution of Building Services, the Confederation of British Industry, the Department of Education and Science, the Department of Health and Social Security, the Institution of Electrical Engineers, the Institution of Mechanical Engineers, the Property Services Agency, the Royal Institute of British Architects, the Royal Society of Chemistry, the Society of Chemical Industry and the United Kingdom Atomic Energy Authority sponsored a Conference in June 1982 entitled *Labdesign 82*. The aim of the Conference was to bring together users, designers, and providers of services of all types of laboratory accommodation in an attempt to exchange the latest information on the design and provision of services for this type of building. The

Table I.2

Progress stages in a laboratory project

Prebriefing
Formation of the project team
Formal briefing
Development of sketch plan
Presentation of the sketch plan and agreement to proceed
Main briefing
Development of full sketch plans, presentation and final agreement to proceed
Production of working drawings
Tendering by contractors
Selection and placing of contracts
Demolition and site preparation
Construction
Removal planning
Fitting out
Phased or total occupation
Rectification of faults and problems during use

---

secondary aim was to provide a forum for both users and designers to exchange information on their problems and difficulties when designing or occupying new or refurbished accommodation.

The chapters in this book are edited up-dated versions of major papers given at that Conference, supplemented by additional material on areas not covered at the meeting. They deal with the issues and problems involved in the provision of laboratory accommodation, provide guidance on good practice, review current methods and consider future developments. The text has been prepared as a contribution to the literature on the design of accommodation for all types of laboratory and to be comprehensible whether or not the reader attended the event. Chapter 1 presents a very personal view of a leading scientist on the problems facing a user when commissioning accommodation. Subsequent chapters provide some but not all the answers to the questions he raises.

R. Lees
A. F. Smith

1

Trends in laboratory work – their implication on the building and conversion of laboratories. A personal view

E. R. Laithwaite
Imperial College of Science and Technology, London

INTRODUCTION

Laboratories have changed their nature dramatically in the last century, at least in industry if not so much in teaching. On the grand scale one can go all the way from elaborate underwater complexes that make James Bond's extravaganzas look almost commonplace, to incredibly 'clean' rooms where research on micro-engineering is carried out.

My own speciality lies in teaching laboratories and particularly in the provision of laboratories for both undergraduate and postgraduate students, plus more elaborate facilities for post-doctoral fellows. In the 1940s this was in only one specialised discipline – engineering. This is no longer true and now electrical engineering would be a more accurate description. In my case, certainly not the whole of electrical engineering, only what is now called 'heavy' electrical engineering. This in turn is divided into electrical machines, and electric power transmission and distribution, and my preserve is only with the former. One is entitled to ask, perhaps, 'Are you then concerned with *all* types of electrical machine?' to which the answer is: 'No, primarily induction motors, and so far as research is concerned, mainly *linear* induction motors'.

THE EXPLOSION OF KNOWLEDGE

The foregoing paragraphs highlight one of the problems, not only of laboratory design, but of almost all aspects of modern life. The explosion of knowledge in science particularly has thrown up so many facets of scientific and technical 'know-how' that many establishments may well find some of their staff fully occupied in re-organising and sub-dividing everything from stores to Directors in an attempt to keep 'with it' or, what is perhaps worse, to be seen to be keeping 'with it'!

It is less than a slight exaggeration to say that in the last two years I have lost by this process two of my most valued and respected senior staff so far as research is concerned. Two years ago they were both good practising 'machines men'. Without warning, one had an instant personal chair in Computer-aided Design (later commuted to Applied Electromagnetics). The other joined a new group – Robotics. What are they going to need in the way of laboratory facilities that they did not require two years ago?

The answer for the first man is fairly easy. He just wants a massive computer terminal with the facility of being connected to a mighty computer just north of Giggleswick, or some similarly unlikely place. But the second man might want anything from merely the same as the first, to a whole mechanical workshop of his own.

THE COMPUTER AGE

In 1951 I was secretary to the Inaugural Conference of the Ferranti Mark I computer, the first full-scale machine to be manufactured and sold commercially and the brainchild of the late Sir Frederic Williams. None of us, I am sure, realised just what kind of a monster we were unleashing on an unsuspecting world. It was not long before a music-hall joke was coined that was to be prophetic. It told of the year 3076 when the whole Universe was, for the first time, under central government and was all to be run by computers. The day came when the Prime Minister of the Universe was to inaugurate the Central Master Computer and to put to it the first question, which fairly naturally was, 'Is there a God?'. After only a fraction of a second's delay, the voice of the computer boomed, 'Yes – there – is – a – God (then a short pause) – now!'.

Computers were quickly followed by pocket calculators and the next century was rapidly being shaped and it was not quite so long before there were some fairly dramatic and disturbing results. Recently, I was in a supermarket with my wife. She chose a piece of frozen fillet steak which was easily the best piece in the freezer, but a bit of the label that stated the price and other particulars had been torn off. When we came to the check-out girl she said in a soul-less voice, 'There's no price on this, would you mind choosing another piece'. I pointed out that what remained on the label was the weight (1.60 lb) and the price per pound (£3.89) and said, 'If you multiply the two you will know the price. Look, I'll do it for you on this old envelope'. She gave me a look as if I was about to steal the Crown Jewels and called over a young man of about 23. 'There's no price on this meat . . .' The process was repeated. The only description of the look in the young man's eyes was *fear*! Is this the shape of things to come? Not only can no-one multiply, no-one knows it is *correct* to multiply.

THEORY TAKES OVER

It is questionable, however, whether the advent of the computer is the main reason why it has become unfashionable to include demonstrations in lectures, whether to undergraduates, in professional institutions or to the public, why more and more emphasis is placed on theory and why such notable establishments as Massachusetts Institute of Technology (MIT) should have scrapped their machines laboratories in the 1960s.

The appearance of the computer coincided with a growing belief in the minds of theoretical physicists that their kind now knew nearly everything, and what was still to be discovered would be done by equations. To be realistic, they had a lot of evidence to back up such a belief. As far back as the last century James Clerk Maxwell predicted the existence of electromagnetic waves in empty space, purely by mathematical manipulation. The tiny bit of matter that disappeared at Bikini Atoll and the resulting devastations at Hiroshima and Nagasaki were the result of theory well done.

But the notion spread, exceedingly fast. From particle physics to chemistry, to biology, some of which was given the fashionable name genetic engineering to add to its power; it soon spread to engineering itself. Gabriel Kron preached worldwide that there was only *one* equation to be solved for all problems in electromagnetism, and that was $E = IZ$ (i.e. the voltage drop across a resistance = current $\times$ resistance). After that, each special case, meaning perhaps the design of a particular alternator, could be treated by matrix partitioning alone. But how many really revolutionary ideas in electrical machines came out of MIT or of any university that worshipped daily at the shrine of the Generalised Electrical Machine? In physics, a great many new things came out of the theory, but engineering is a quite different kind of discipline where, I maintain, there is no substitute for the experimental result.

At school one is taught to solve n equations for n unknowns. The research physicist, armed with a computer terminal, can solve n equations for $(n + 2)$ unknowns, using 'hill-climbing' and other sophisticated techniques to optimise the values of the two extra unknowns, effectively without having to try all values. Perhaps for the first time, in the 1980s, the physicist has accepted that $2 \times 2 = 3.99$ which rounds off to 4, but the engineer has known this since before the Pharaohs!

So new schools were built and their laboratories had the traditional sinks, water taps and gas taps and only 230 V – single phase at that – on the teacher's bench. The Examining Boards ploughed on through the 1960s and into the

1970s, setting questions in practical physics on the magnetometer, and calorimetry, mostly because they knew the schools had all got the apparatus, and there were many justifications for doing this.

'BIG' TAKES OVER

One of several popular misconceptions bandied about in the 1950s and 1960s, especially in Heavy Engineering was that only huge teams of workers with hundreds of millions of dollars, or roubles, behind them could contribute to *real* research. Anyone else was just playing with his or her Meccano set. The fact that this justified the designers of school laboratories making only minor changes from the Victorian idea of a school laboratory met with approval all round from those who had to foot the bills.

Without doubt, the way to justify one's existence in research in the 1970s was to think BIG, and it was done no better than at over-staffed Government research laboratories, especially in pure physics. When invited to give a lecture at one such establishment, I was fêted by having the Director give me a personally conducted tour of the new laser laboratory. With the demise of the ill-fated Tracked Hovercraft still half choking me with disgust, I remember assessing the cost of bending that laser around each right angle and reckoning up the cost of that laboratory, not as so many millions of pounds but as so many Tracked Hovercrafts (which had cost £5.25 million at the time of closure).

At the end of the tour I saw a wall that apparently contained a rectangular hole about 60 X 30 cm over which had been fixed a piece of hardboard. 'This', said the Director proudly, 'is where the beam finally emerges from the "clean room" into the main laboratory outside'. 'And what happens to it then?' I asked. 'Well, then it's available for anyone who wants to use it,' was the amazing reply. I was reminded forcibly of an episode in the wartime radio show *Much Binding in the Marsh* where an actor, Wallace Eaton, always had an original and even more pointless job, one of which was to put the punch holes back in bus tickets. When asked what happened once he had found one that fitted he replied simply, 'Well that one's been checked!'

Given the chance to go back to 1966 and start the Tracked Hovercraft project again I would have made every effort to make sure they had spent £100 million by 1973, then they dare not have closed it! Tracked Hovercraft was certainly a BIG project, but obviously not big enough to survive.

THE ACCOUNTANT TAKES OVER

The biggest take-over of all was by those who controlled the money. The tip of the iceberg appeared in the 1930s when such isolated examples as one I experienced in Manchester occurred. An employee in a multiple store got a bonus for life in his pay packet for an idea dropped into the Suggestions Box. The store sold goldfish in bowls, on the ground floor. He suggested that if, instead, they were sold on the fourth floor, the likelihood of the bowl being broken in the jostling on the stairs was quite high. No-one but a monster would leave a fish struggling for life on the steps, to be trodden on and anyway it was little Johnny's fish and he would scream the place down if another bowl was not bought immediately. The sale of bowls increased by 30 per cent on a busy Saturday!

During the Second World War, the Americans brought over the idea of 'built-in obsolescence', a phrase coined to make all exercises such as the goldfish bowls look at least faintly respectable. The age of the Accountant had begun. Accountancy began to dominate almost all facets of life, except the exotic. The accountant is with us and, no doubt, will tell us that if laboratories are designed in this or that way, they could be converted easily to growing mushrooms in the event of a top-level change of policy on defence, with the minimum interference of cash flow. The result of such exercises is often a poor design for either purpose.

In academic life the effect of the accountant has been strange indeed. Whether advising the University Grants Committee or the Science Research Council or the foundations and trusts, the answer was the same. There had to be a father figure somewhere. You can have an electron microscope or a personal computer for a million pounds, if you make a good case, but you can never, never have a technician to service it. You may be given a personal chair in Group Dynamics or some similarly obscure title but you may not have a personal secretary.

THE HUMAN ASPECTS

What the planners, financiers and designers perhaps miss most is that research workers are human. A lot of these workers prefer snacks to a four-course lunch and they would like somewhere nice but very close at hand in which to eat them. They are very fond of personal encouragement in the form of cash for giving their all to a project. They would like a say in the design of any new laboratory in which they are to work and they would like their voices to be louder than that of the architect on many occasions. They dislike being taken off a project just when it is becoming interesting and in this last connection I would quote a well-known Director of the Kodak Company, Mr C. K. Mees, on the subject of what research should be done:

> The best person to decide what research is to be done is the man doing it, and the next best is the department head. After that, there are only increasingly worse groups — the Research Director who is wrong more than half the time, a Committee, which is wrong most of the time, and finally a Committee of vice-presidents, which is wrong all the time.

I thought that this criterion might well be applied to the design of laboratories, for if you reverse the order of the list, apart from the vice-presidents, my guess is that you would have a measure of the extent of the influence that people have in such designs.

Section 1:

DESIGN, COSTING AND CONSTRUCTION

2

Design trends – implications on building and converting laboratories

J. Weeks
Llewelyn-Davies Weeks, London

INTRODUCTION

Before starting to design a building, architects always ask their clients for a very detailed brief since it seems axiomatic that if this is very perfect and complete, the client will have a very perfect building. It is an unfortunate truth, however, that in laboratory design, a very perfect brief, if complied with in every detail by the architect, will result in an obsolescent facility. In the three or more years between the time the architect gets the brief, and the scientist gets the laboratory, the work will have changed, the equipment will have changed, and many of the individuals who prepared the brief will have dispersed. However, the old tenet on which architects are usually educated – 'form follows function' – is fortunately not true. If it were true most of the buildings in use today would be extremely uncomfortable and indeed in many ways impossible to use, since very few are used as they were intended. Our building stock is fixed, but fortunately human beings are very flexible and adjust themselves to the environment they find. If architects ride this lucky break too far they are avoiding their duty to society. Society looks to them to assist it to live its life, not to impede it.

Llewelyn-Davies Weeks has designed many laboratories for medical research, and like all research this is, by definition, unpredictable; there is therefore the problem of reconciling the unpredictability of the programmes with the long-term usefulness of the building. Buildings are appallingly permanent but research programmes are relatively ephemeral.

This chapter will consider examples of laboratory buildings, together with some theoretical background. From these will be drawn the essential design requirements for a laboratory which will impede the life of the laboratory users as little as possible.

THE ARCHITECT'S VIEW OF A LABORATORY

Architects looking at a typical laboratory in use see something which is very different from anything which they want to see in their buildings. Architects are obsessive to a degree – in an ideal situation they will design the cups, choose the curtains and keep everything tidy. The architect may wonder what is to be done about the typical laboratory worker's apparent complete disregard for the elegance of the environment. In the event, of course, the architect must discover the required infrastructure, the hard bones of the laboratory which can support the untidy events of everyday work with some kind of convenient order.

An empty space, a building shell, can be a laboratory if the right services can be provided. A laboratory, a hospital and an office building can all utilise the same basic building shell if the shell has certain characteristics. Essentially, the structural shell of a building will last longer than any other part. To get the longest use from it, therefore, it has to be designed to accommodate change in the short-life components, in particular the services. A correctly designed permanent shell will allow everything which it supports and encloses to be impermanent. For research laboratories this is the greatest facility a building can offer, for given enough weather-protected space and accessible services any kind of laboratory work is possible.

It is useful for an architect to remember that the things which make up a laboratory building do not all have the same lifespan. The concrete is there for life. Partitions are impermanent, and in any modern framed building they can be altered easily. Services have a shorter lifespan than partitions; they will certainly be changed as the nature of the work changes. The benches must be movable at very short notice and it should not be necessary to bring in a work crew to move them; it should be possible, to move them about like tables. Lastly, doors are like instantly movable partitions – a locked door is a wall, an unlocked door is an opening.

Laboratory planning is based on a small set of dimensions which are derived ergonomically from people at work. Some of these are illustrated in Fig. 2.1. The dimensions shown are not precise but an approximation. The work top is 600 mm deep, a typical measure of a conveniently outstretched arm. Behind the work top is a 150 mm zone for piped and wired services, water and gas supply, and drainage. If two people are working back to back and somebody is walking between them, a dimension between the walls of more than 3 m but less than 4 m is required to give comfortable space for everyone. This dimension is the width of an element of working space from which all other working dimensions can be derived.

The number of dimensions required can be rationalised on the basis of 7.2 m, that is twice the dimension of the fundamental element of 3.6 m. 7.2 m, being duodecimal, is neatly subdivisible arithmetically; at the lower end are dimensions which relate closely to the familiar Imperial dimensions 4 in, 6 in and 1 ft. Some of these dimensions are listed in Table 2.1. A building based on a

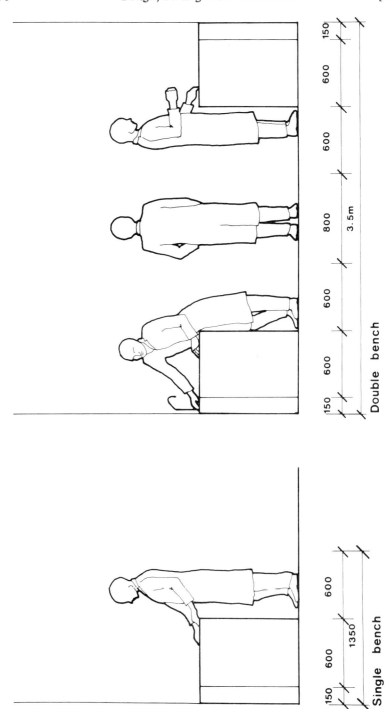

Fig. 2.1 – Ergonomic dimensions.

Table 2.1

Dimensions derived from 7.2 m

Arithmetic division	Linear dimension (m)	Examples of applicability
1	7.2	Structural module Main services centres Full laboratory module
2	3.6	2- to 4-person laboratory Electrical risers Office Special procedure room
3	2.4	Writing space Fluorescent tube (2.44 m) Small office Special procedure room 'U'-shaped work layout
4	1.8	Writing space Fluorescent tube (1.83 m) 'L'-shaped work layout
6	1.2	Modular work top Fluorescent tube (1.22 m) Safety cabinet Wide door, including frame
8	0.9	Domestic door, including frame

dimensional grid of 7.2 m will have a modular base which will accommodate everything required; the structure and services reticulation at the upper end, and the tables, cupboards and doors at the lower end can all relate to the same dimensional discipline.

In summary, the kind of building shell needed is one in which the structure is separated from the services and will accept changes in a flexible way, and designed around dimensions which are based on ergonomic necessity.

EXAMPLES OF DESIGN

1. Greenwich Hospital

What kind of building results from these rather simple theoretical requirements? Figure 2.2 shows a section through a hospital designed by the Department of Health and Social Security and built at Greenwich some years ago. It introduced into its structural frame what is known as interstitial space, that is a space covering the whole building at every floor, between floor and ceiling. All the service distribution runs are in this space. The lower deck is not just a suspended ceiling, it is a floor and it provides a working surface for maintenance crews for the whole services system. Long lattice beams span the building, and the services run through the voids in the lattice. There are no columns. Services can be fed through the ceiling to rooms below at any point, and the flexibility is only limited by the dimensions of the ceiling panels. Unfortunately the depth of this interstitial space may be 2.5 m, or more, and this extra building volume has to be paid for on each floor of a building which is serviced on this system.

This system is very elegant, but can it be afforded, and is it an essential requirement? It can be afforded more easily if the building has to be air-conditioned, but if there is no imperative requirement for this, climatic, for example, or to alleviate noise or urban pollution, it is rather an expensive way of servicing. Undoubtedly it is flexible, but the question to be asked is whether the high degree of flexibility conferred is worth the money spent.

In common experience, even where there is the possibility of placing partition walls anywhere and tapping services to match, the room sizes used are not inconsistent with a more rigid planning discipline. It is possible to take the view that interstitial service floors are a romantic reaction to the quest for planning flexibility. If the distribution of different room sizes is observed carefully, the degree of flexibility required to accommodate different work regimes can be seen to be provided within the relatively small number of ergonomically derived dimensions [1].

2. Nuffield Institute of Comparative Medicine Laboratory, London Zoo

Twenty-five years ago, a laboratory was built at the London Zoo for the Nuffield Institute of Comparative Medicine to a design by Llewelyn-Davies Weeks. A floor plan is shown in Fig. 2.3. There are regular vertical service ducts between structural columns, at 3.6 m centres. These ducts, each with doors, are provided on both sides of the corridors so that all the services are accessible from the corridors; no services are buried at any point. The services for the individual laboratories run out, behind the benches, from these vertical shafts along the partitions, in demountable spines. Each laboratory is about 3.6 × 7.2 m, and sometimes a double module is used, or a module may be subdivided into smaller rooms. All benches are mobile and all partitions can be moved. Cross partitions can provide offices by the windows and internal special environment rooms which can be serviced from the main ducts as are the full-depth laboratories. The central core is air-conditioned.

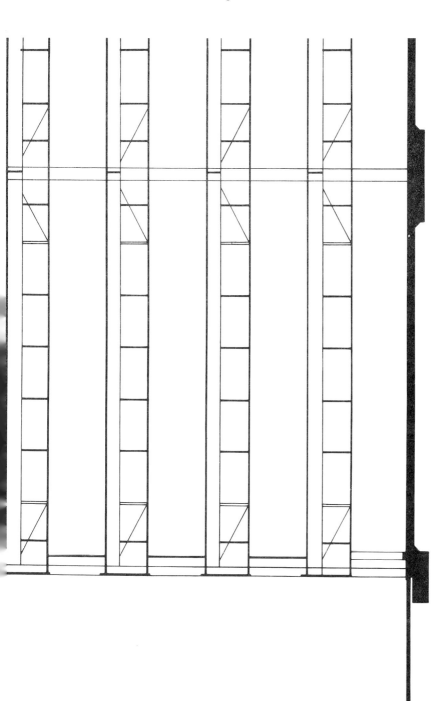

Fig. 2.2 — Section through a building with interstitial service space.

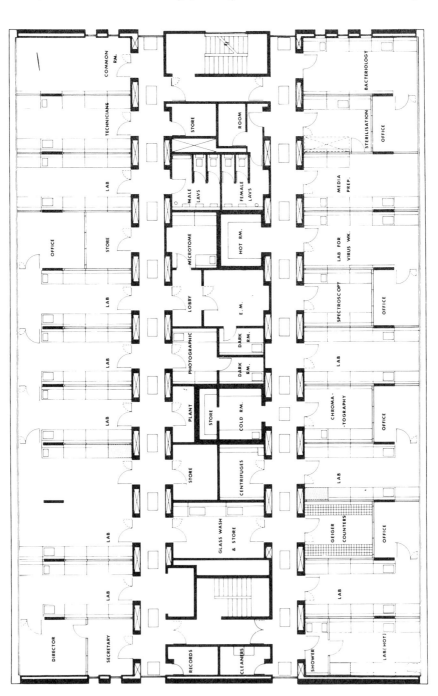

FIRST FLOOR PLAN

Fig. 2.3 – Nuffield Institute of Comparative Medicine Laboratory: floor plan.

While the plans show many service ducts, this has to be compared with the spaces provided in the building with interstitial floors (Fig. 2.2). The degree of flexibility provided in the Zoo building is very high for services, and for room sizes. Is more flexibility required?

3. Clinical Research Centre, Northwick Park Hospital
Figure 2.4 shows the plan of the Clinical Research Centre at Northwick Park, Middlesex, built between 1965 and 1975. The Centre houses a very wide range of work, both clinical and non-clinical. The services ducts are, as in Fig. 2.3, at 3.6 m intervals. The degree of flexibility provided, of which the users have taken advantage, has been satisfactory.

4. Clinical Research Building, National Hospital for Nervous Diseases
The plan of the clinical research building at the National Hospital, Queen Square, London, is shown in Fig. 2.5. This is a largely air-conditioned building, and the vertical ducts provide alternately air-conditioning distribution and piped services, so that all four walls of each laboratory module can be serviced.

SERVICES

In the laboratories and buildings described above, the electrical outlet spines are separated from the water and gas services, 300 mm above the work tops, so that they are more easily accessible. The work benches are movable tables with mobile under-bench cupboard and drawer units. The latter have wheels at the back only and can be moved very easily. The ventilation trunking is exposed so that fume hoods can be connected where required.

The Nuffield study on laboratory planning [2] was carried out at a time when, in laboratories, there was still a great deal of wet bench work, only a relatively small amount of work using electronics, and no computers. As electronic work has grown in volume and wet bench work diminished, it has been found possible to omit alternate piped services ducts. Each module has one long wall available for work using gases and water, while the other long wall and the window wall is serviced with electricity and communication lines only. Very often, now, double modules (7.2 X 7.2 m) are used in which ceiling-hung pendants for electric outlets replace the central partition. Along the window wall there is a bench for writing, with electricity and communication lines.

ROOM CONFIGURATIONS

Figure 2.6 shows a building shell and five alternative configurations of rooms for scientific work. This has been designed for a mixed clinical and research, high-rise building in Hong Kong.

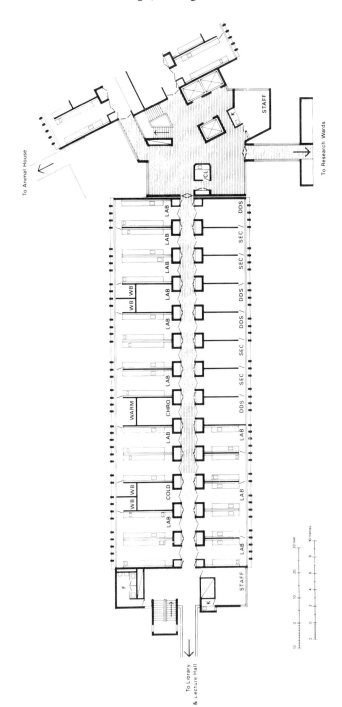

Fig. 2.4 – Medical research laboratories at Clinical Research Centre, Northwick Park Hospital. Key: CL – cleaner; CHRO – chromatography; COLD – cold room; DDS – Divisional Director's study; F – female cloakroom; K – kitchen; LAB – laboratory; SEC – secretary; STAFF – staff rest room; WB – writing bay; WARM – warm room.

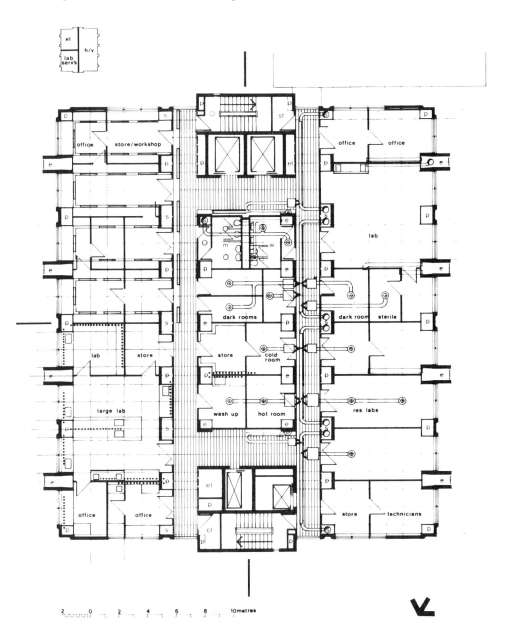

Fig. 2.5 – Clinical research building at National Hospital for
Nervous Diseases, Queen Square, London.

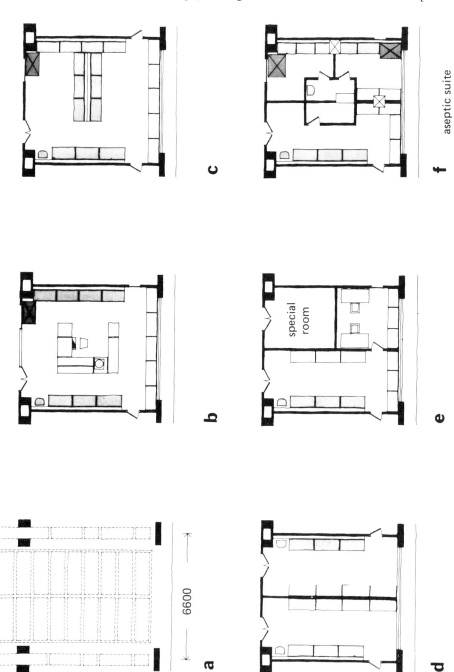

Fig. 2.6 – Variations of room configuration within a fixed structural shell. For explanation see pp. 41 and 45.

The structural components are shown in Fig. 2.6(a). The structural grid is 6.6 × 7.2 m, with the interior columns paired. Dotted lines indicate the extent of perforations which are possible through the floor slabs.

A typical laboratory using the full structural module is shown in Fig. 2.6(b). All services are provided at the side walls, while electricity only is provided at the centre from ceiling pendants which may be arranged freely. The run of tables under the window has electricity and communication lines.

Figure 2.6(c) shows a laboratory arranged with a peninsula working surface with all services. In Fig. 2.6(b) and (c) a site for safety cabinets is shown in a corner where it is protected from draughts.

Two half-module laboratories, each with full services on one wall and electricity and communication services on another, are shown in Fig. 2.6(d).

Figure 2.6(e) shows one half-module as a laboratory with the other divided into an office at the window and a special environment room at the corridor side of the module.

A special suite for aseptic work planned within the standard shell is shown in Fig. 2.6(f).

Many other configurations are possible.

AN ALTERNATIVE SERVICING SCHEME

One other system of laboratory servicing is of interest. During the mid-1970s the Department of Education and Science set up a study group to consider laboratory designs for schools and universities. In the laboratories which have been designed following the study, a planning grid is implied by a regular arrangment of permanently installed drainage points in the floor of each laboratory area. These are used where required, and blanked off where not. All other services are suspended from the ceiling and run out from comparatively rare vertical mains distribution ducts. Bench services are tapped from the ceiling distributors where required and brought down in flexible leads to working level where they are connected into a services bollard. This supplies gases, electricity and water and is located at a convenient place on the work top. Sinks are built into work tops, or stand as free-standing items on top of low tables, and are connected to the nearest floor drainage point by flexible pipes. Pilot projects have been built and a careful note made of how the work tops have been moved as the users found it necessary to modify the original layout.

Clearly this arrangment is highly flexible and is not dependent, as is the system which uses interstitial spaces in the structure of the building, on an inherently expensive building form. Columns can be used at normal centres, and between the columns a high level of flexibility is available. Since all services are supended from the ceiling they are exposed, likely to collect dust and the ceiling appears cluttered; all benches are connected to the ceiling by the umbilical service cords. A criticism of this system therefore is that the laboratory starts life by looking untidy. When the usual clutter which arises automatically from laboratory work is added to this basic clutter, the interior of the laboratory seems to be very busy. This is an aesthetic view and may not be important if the laboratories, in use, prove to be as serviceable as they set out to be.

A variation on the system, which is being used experimentally, has drainage lines also suspended from the ceiling. Small pumps operate automatically at intervals to pump the sink effluent from catch pots into the overhead distribution system. Using this system, no perforation of the floor is required for drainage, and an additional component of flexibility is added which allows any space in any building to be converted for laboratory purposes. Upward drainage being dependent on a mechanical pumping system is inherently less reliable than gravity. Nevertheless, it is claimed that the system does not break down and that preventive maintenance and cleaning are easily managed on a routine basis. The ability to convert any existing room into a fully serviced laboratory by use of the components of this system is an important asset and, when finance for new buildings is difficult to arrange, a particularly significant one.

REFERENCES

[1] J. Weeks, G. Best, J. Cheyne and E. Leopold, Distribution of Room Size in Hospitals, *Health Serv. Res.,* Chicago, 1976, **11**.
[2] The Nuffield Foundation Division for Architectural Studies, *The Design of Research Laboratories,* Oxford University Press, London, 1961.

3

Flexibility and adaptabilility – alternative design strategies

R. E. Clynes and A. J. Branton
Laboratories Investigation Unit, Department of Education and Science

INTRODUCTION

The Laboratories Investigation Unit (LIU) based at the Department of Education and Science (DES), advises on laboratories for teaching and research at all levels of education, as well as those for industry and health, in the United Kingdom and overseas. Its connection with DES ensures that the Unit is particularly conscious of the need for economy while meeting new requirements, including those resulting from legislation, and for achieving value for money.

In the area of education the LIU is faced with schools which lack basic science facilities, and with those which need a temporary upgrading of existing poor facilities during reorganisation of an area. The Unit is faced with polytechnics and colleges which sometimes have make-do facilities in leased buildings and with universities having, for example, old fume cupboard installations which do not meet the current needs of health and safety, and are housed in buildings where the structural and planning restrictions are severe. These factors are not limited to education and have influenced the focus of recent work of the Unit. They will be discussed later. The material of this chapter draws on a forthcoming book[1] by one of the authors.

FLEXIBILITY AND ADAPTABILITY

The first lesson learned by the LIU when it was set up in the late 1960s was that requirements within the laboratory are bound to change. These changes are often unpredictable and in many cases frequent and they must be met as best they can. The message carried through the Unit's development projects and publications has been that accommodation should as far as possible be designed to be flexible or adapatable in use. The aim has been to cater for the cases where future predictions are found to have been wrong.

The terms 'flexibility' and 'adapatability' are commonly used in a loose interchangeable way to imply that the accommodation allows a degree of responsiveness to changing user needs. The following definitions for these terms will perhaps help to clarify the different emphasis in the approaches to design that can be developed to meet both initial and future requirements:

Flexibility enables different activities to be accommodated in given spaces without physical rearrangment having to take place.

Adaptability refers to a building which allows physical rearrangement of building elements, services and furniture.

In practice most recent laboratory buildings combine both attributes, but with the stress placed on one aspect or the other.

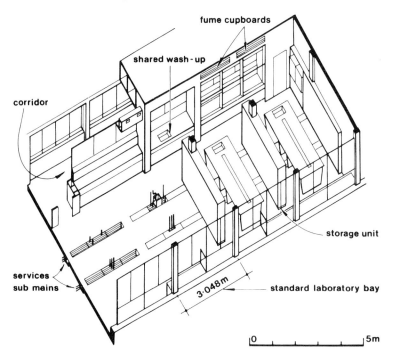

Fig. 3.1 – A very early example of the modular repetitive bay at the research laboratories of the Dyestuffs Division, ICI, Blackley, designed by Serge Chermayeff in 1938.

THE DESIGN APPROACH

Various approaches have been adopted by designers to avoid producing a design which is too constrained by initial needs. These include:

(a) Laying out laboratories on a modular repetitive bay, each unit having a standard pattern of benching and services[2] (Fig. 3.1). The stress in this approach is on flexibility.

(b) Provision of services on a regular grid, in excess of immediate needs, and housed in vertical ducts, floor ducts or from overhead so that any work station is able to tap off the full range offered[2] (Fig. 3.2). The stress here is also on flexibility.

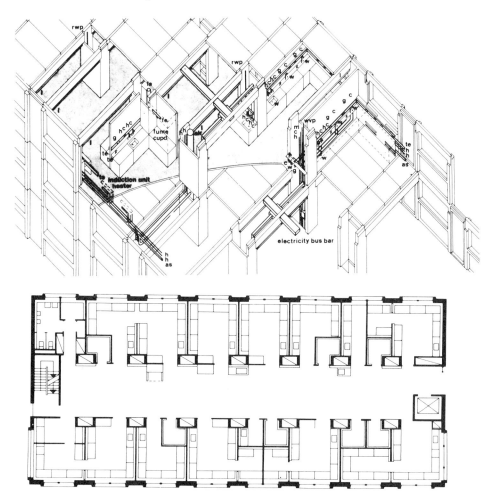

Fig. 3.2 — Distribution of services to laboratory benching from repetitive vertical submain ducting. Department of Zoology Research Building, Edinburgh University by Thomas Kenney, ARU Architects. Key: h — hot water; c — cold water; m — main; w — waste; g — gas; rwp — rainwater, wvp — waste vent pipe; te — telephone and electrics; fe — fume extract; l — lighting trunking; h — heating pipes; as — air supply.

(c) Developing the structure of the building to take a wide range of alternative partition layouts and using removable partitions which can be non-load-bearing brick, blockwork or demountable[3] (Fig. 3.3). In this approach the stress is on adaptability.

(d) Choosing or designing unit laboratory furniture so that it can be added to, subtracted or rearranged as required[3] (Fig. 3.4). The stress here is on adaptability.

Fig. 3.3 – Removable partitioning system which can be relocated to meet different planning needs. The Charles Darwin Building, Bristol Polytechnic by the Laboratories Investigation Unit.

With a flexible design an assessment is made at design stage of the widest likely range of work to take place in the foreseeable future and a generalised, but fixed, arrangement is made. New needs and organisational changes are met by moving people and their equipment rather than by making physical alterations to the layout. An adaptable design allows tailored accommodation for each changing need by making physical adjustment of the facilities. The way in which this is done will vary depending upon how easy it is to make the adjustment.

Fig. 3.4 – Relocatable unit furniture serviced from overhead for the School of Sciences, Trent Polytechnic, Clifton by the Nottinghamshire County Architects Department.

THE PROVISION OF NEW FACILITIES

The modular approach has tended to equate new projects with new buildings. The period 1960 to 1970 was a time of growth for new laboratory buildings. This is no longer the automatic way in which new facilities are provided.

Apart from the custom designed new building, there are many other ways. Existing laboratories may be renewed, an 'off-the-peg' total package may be provided, or an owned building converted for change of use. Other possibilities are the provision of an 'off-the-peg' new building shell with a custom-designed interior, the use of leased accommodation, or, to meet short-term needs, the use of easily transportable facilities.

These alternatives do not necessarily preclude a modular approach, but clearly any space must set its own limits on what can be done. In older-established science complexes in higher education, government research or private industry, it is not uncommon to see facilities provided by two, three or even more of these alternatives. The important factor with any new project is to analyse carefully what exists and assess what approach will be most applicable for the project with the aim of working towards maximum interchangeability within and between buildings.

NEW BUILDINGS OR REFURBISHMENT?

When faced with the apparent need for new accommodation, it is important to question all initial assumptions before embarking on a building project with all that entails. This may be illustrated by two projects in which LIU was involved.

In the late 1970s Wye College foresaw a lack of teaching space as the number of its students grew. The initial, and perhaps instictive, reaction was to think in terms of building more accommodation. If this had been the only approach, the College would have had to find the space and capital to enlarge its estate and also the extra recurrent cost of running it.

Students at Wye College have a choice not only between combinations of courses but also of options within them, making timetabling complex and more than normally difficult when done manually. The considerable variation in time-tabling resulted in the teaching laboratories being under-utilised. LIU proposals [4] showed that by modifying the existing laboratories and, in particular, by providing two multi-discipline laboratories each capable of housing any practical subject except geology, utilisation could be improved to the extent that the College would not need new accommodation, but could also release some of its existing laboratories for other purposes. DES had available a computerised system for timetabling and space allocation. After a successful trial this was adopted by Wye College.

A similar feasibility study was carried out for Du Pont (UK) Ltd who assumed they would have to extend one laboratory building when a second laboratory complex in a process building was closed down. Reorganisation allowed better grouping of shared facilities and a higher level of use and adaptability which generally proved to meet all the firm's foreseeable needs without extending the building.

RENEWAL OF EXISTING LABORATORIES

The arguments for adaptable laboratories in new buildings are no less valid in the context of renewed accommodation. Exchanging one fixed installation for another makes little economic sense when related to the changing needs of a developing organisation.

The Biochemistry Department at the University of Cambridge occupies a 1924 listed building. The teaching laboratory on the top floor of the building needed to be modernised. It had fixed benching, original wiring, open channel drains and service ducts in the floors. Additionally, the roof lights leaked and gave solar gain and the laboratory was cold in winter.

Following an LIU feasibility study, the work was dealt with by the University Estates Department[5]. The aim was to provide maximum adaptability within the constraints imposed by the existing structure. An open plan area was provided which allowed for changing research needs and the relocation of equipment. Some small permanent spaces, mainly relating to structural columns were also provided. These served as areas for switchgear, balance room, centrifuge bay, cold room, small sterile room, and chromatography room. Demountable office units each with its own ceiling and double glazing were incorporated in the scheme.

Since its completion in June 1980 this conversion has presented very few problems. There is complete flexibility in services. Gas, electricity, hot and cold water and compressed air are carried in overhead booms. Spine units carry the bench services and the bench tables, bolted to the spine units as required, are left completely free from services (Fig. 3.5). This system was similar to that used at Bristol Polytechnic[3]. A stock of spare fittings was supplied in the original contract; this allows for minor relocation of equipment to be made as necessary if work programmes change.

The Queen Elizabeth II Hospital in Welwyn Garden City provides another example of a project for renewal of existing laboratories. The Pathology Department of the hospital is located on the second floor of one wing of a building constructed during the 1960s. The floor is shared with the Central Sterile Supplies Department. It became clear that there was a need to expand the Pathology Department and improve facilities in order to meet the requirements of the Code of Practice on the prevention of infection in clinical laboratories[6] which resulted from the report of the Howie Committee.

A survey showed that the corridors were dangerously overcrowded and rooms were divided by a mixture of fixed and mobile partitions. The services were on the perimeter of the building and in the case of vertical services these were not always in the places shown on the plans.

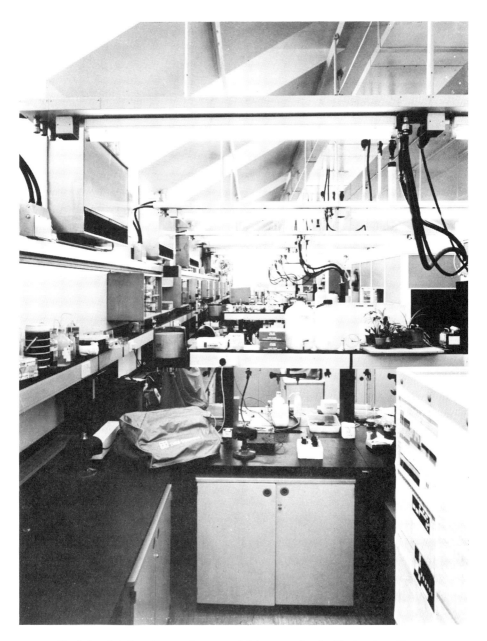

Fig. 3.5 – Interior of open-plan research laboratory for the Biochemistry Department, Cambridge University by the University Estate Management and Building Service.

The LIU proposals recommended that the laboratories should be grouped together so that they could be rebalanced more readily; offices also should be grouped together. New partitions should be used which were free of services and fittings, and sub-division of space achieved using screens or demountable partitions. The main service runs would be carried on overhead booms with drainage systems on the perimeter of the building. Loose furniture would be used in both laboratories and offices. The estimated cost for this work was £100,000.

'OFF-THE-PEG' PACKAGES

'Off-the-peg' packages are becoming increasingly available where part new accommodation is required. These avoid the need to become involved in the complexities and time delays which can result from setting up a new building project. A typical package is that marketed by Tec Quipment Ltd in which self-contained unit laboratories can be joined together and grouped as required. These units can also include a controlled environmental facility. Another type is the MSS Standard containerised animal house packages which cover a wide range of units for medical and scientific work.

CHANGE OF USE OF AN EXISTING BUILDING

In 1975, during the course of regrouping from fourteen sites to five, Middlesex Polytechnic bought a speculatively built office block of 2600 m² and a factory/warehouse of about 11 000 m² at Bounds Green. The architects employed to carry out the adaptation of these buildings incorporated a clear pattern of main and secondary circulation routes and added mezzanine floors to give sufficient floor space for 1500 students (Fig. 3.6). The existing structure was exploited using exposed services to allow maximum adaptability of the working spaces. Work on the buildings was phased as money became available. During Phases 1 and 2, light laboratories were constructed. Special laboratories, some of which were cooled, were included in the later phases of the conversion. This scheme showed a flexible approach to education, the use of an adapatable plan and the development of shared facilities wherever possible.

THE USE OF LEASED BUILDINGS

A recent approach in speculative development is where accommodation is designed specially for letting. This is particularly suitable for small firms who are setting up in business. One example of this type of development is the Genesis Building at the Birchwood Science Park in Lancashire. This was designed by the Warrington Development Corporation for leasing as laboratory space to small companies. Once established, a firm would move to larger premises of its

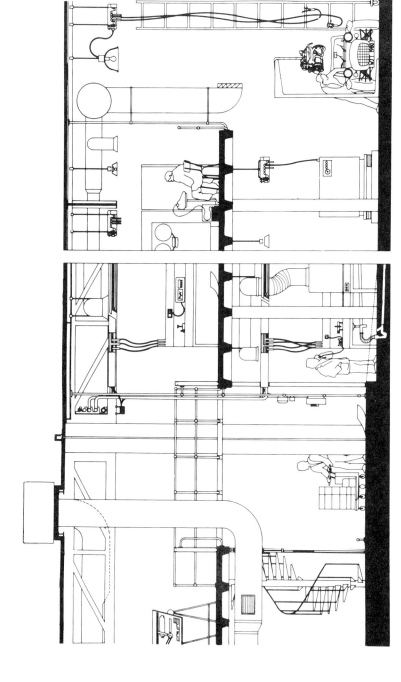

Fig. 3.6 — Cross section showing the conversion of existing factory space to adaptable laboratory and workshop facilities for Middlesex Polytechnic by Martec International.

own and another user would take over the vacated premises. The designer has therefore to incorporate a high degree of adaptability (Fig. 3.7).

A similar development at the Aston Science Park (Fig. 3.8) is a joint venture by Birmingham City, Aston University and Lloyds Bank. The holding company is converting an existing and extensive factory for this purpose. A completed block is to be used for office-type accommodation and a larger one, which will include laboratories, is due to be completed early in 1983.

In education a more common approach is where the user institution looks around for suitable accommodation which is then fitted out to suit its needs. An example of this type of development is Central House which is a speculatively built flatted factory on a near-island site where atmospheric pollution, traffic

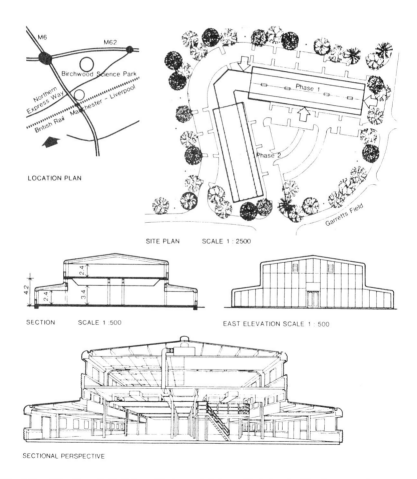

Fig. 3.7 – The Genesis Building, Birchwood Science Park by the Warrington and Runcorn Development Corporation, and DEGW Space Planners.

(a)

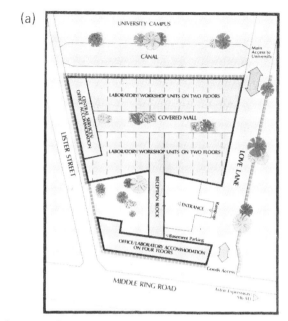

(b)

Fig. 3.8 – The Aston Science Park Building by the City of Birmingham Architects Department. (a) the plan showing the possible subdivision of space into lettable units for laboratory or workshop use. (b) a view of the covered mall giving access to the laboratory/workshop units.

noise and solar gain are all acute. The Materials Science Department of the City of London Polytechnic was poorly housed in part of this building on which less than 30 years of the lease remained. The Polytechnic was faced with the choice of either undertaking an expensive renovation and possibly having to write off its cost after a short period, or to do as little as possible on the building. The latter would be unlikely to provide a decent working environment for the users. Neither solution would have represented good value. There was an added complication in that the future of this department in the Polytechnic was uncertain pending the result of a review by the Inner London Education Authority (ILEA).

An LIU proposal[7] showed how the accommodation, which had grown piecemeal, could be rearranged to give more efficient zoning and working areas for the Materials Science Department. The interior could also be fitted out so that it would be equally suitable for whatever discipline might be moved there, as most of the interior systems would be able to be moved elsewhere if necessary (Fig. 3.9). The main features of the proposal were:

(a) a suspended ceiling grid to carry electrical circuits; this allowed flexibility;

(a)

Fig. 3.9a – Laboratories for the Materials Science Department, City of London Polytechnic. A view of the leased flatted factory building in Whitechapel.

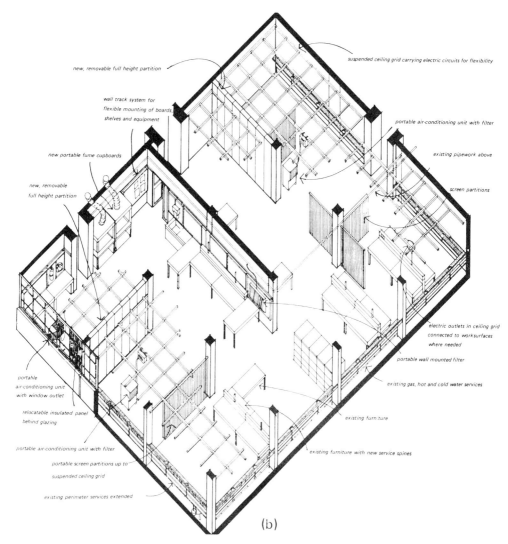

new, removable full height partition

wall track system for
flexible mounting of boards,
shelves and equipment

new portable fume cupboards

new, removable
full height partition

portable
air-conditioning unit
with window outlet

relocatable insulated panel
behind glazing

portable air-conditioning unit with filter

portable screen partitions up to
suspended ceiling grid

existing perimeter services extended

suspended ceiling grid carrying electric circuits for flexibility

portable air-conditioning unit with filter

existing pipework above

screen partitions

electric outlets in ceiling grid
connected to worksurfaces
where needed

portable wall mounted filter

existing gas, hot and cold water services

existing furniture

existing furniture with new service spines

(b)

Fig. 3.9b – Laboratories for the Materials Science Department, City of London
Polytechnic. Typical interior showing the introduction of adaptable systems
components.

(b) new partitions to be demountable;

(c) use of a wall track system for the suspension of boards, shelves and equipment;

(d) movable air-conditioning units, which would be either bought or hired, to
be installed where necessary; and

(e) mobile fume cupboards to be used.

Shortly after the LIU report was accepted by the Polytechnic, and before work was put in hand, the Polytechnic was offered use of a sizeable group of school buildings in Shoreditch that were to become redundant by 1984. The LIU has again been commissioned by the Polytechnic to look at the implications of moving some science work to these buildings. This is one of a number of options being considered by the ILEA.

USE OF EASILY TRANSPORTABLE FACILITIES TO MEET SHORT-TERM NEEDS

As a means for improving flexibility, LIU has developed a number of 'Labkits' for use in educational laboratories. These include a trolley, a work station and a mobile fume cabinet (Fig. 3.10).

CHOICES FOR MANAGEMENT, DESIGNERS AND USERS

It would be too simplistic to relate flexibility and adaptability to the type of institution or laboratory: the repetitive bay to bench-scale research and medical laboratories, and the adaptable basic/supplementary approach to educational laboratories. The key issue for each project is what combination of flexibility and adaptability is required for the future.

Where groups of large-scale floor-mounted equipment are needed, then an open-ended approach to laboratory layouts is generally adopted out of practical necessity, as the size of such equipment varies so much. The debate therefore centres on accommodation for bench-scale work. The impact of equipment on laboratories is of course likely to continue to grow, although there is increasing sophistication and miniaturisation as well as just more equipment. Careful analysis of work in teaching, research and routine laboratories has shown that certain sizes of space will meet a wide range of needs. Furthermore it is clear that some designers continue to offer the use of standard bays within a given range of room sizes. Other designers believe that all users, present and future, should be offered some personal choice in how their work space is planned.

The balance between flexibility and adaptability will also be strongly influenced by a client's attitude towards organisation. To standardise layouts is a useful compromise; everyone gets the same and there are fewer arguments if a reasonable compromise is made. A more open-ended choice of layout offers a greater challenge. Departments and groups can develop layouts on a more individual and democratic basis but the effective co-ordination of use of proposals, as requirements change, remains the key to a successful use of the adaptable laboratory.

The freedom to replan as well as reallocate the use of laboratory space brings with it the need for readily understood documentation on how to adapt space in a responsible way. Clients and designers need not only to devise simple

(a)

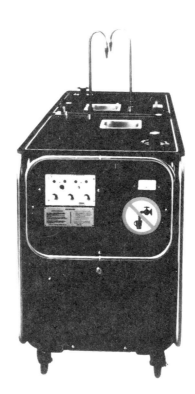

(b)

(c)

Fig. 3.10 – Laboratory fittings designed to be quickly installed and removed to meet short term needs: (a) The Labkit Trolley System for lower school use; (b) The Labkit Workstation System for upper school and further or higher education use; (c) A mobile fume cabinet incorporating filters.

codes of practice, but also to develop foolproof ways of making sure they are used. In this way the laboratory will be a resource able to respond to the evolution of the work done in it. If in the process buildings and accommodation can be provided that lift the spirit a little, then so much the better.

REFERENCES

[1] Tony Branton, *Laboratories: The Design of Facilities for Science and Technology,* The Architectural Press Ltd, in preparation.

[2] *Design of Research Laboratories,* The Nuffield Foundation, Oxford University Press, Oxford, 1962.

[3] *The Charles Darwin Building, Bristol Polytechnic,* Laboratories Investigation Unit, Paper No. 9, Department of Education and Science, HMSO, London, 1977.

[4] *Better Use by Sharing,* Laboratories Investigation Unit, Paper No. 11, Department of Education and Science, HMSO, London, 1981.

[5] Tony Branton and M. J. Purvis, *Lab. Pract.,* 1981, **30**.

[6] *Code of Practice for the Prevention of Infection in Clinical Laboratories and Post Mortem Rooms,* Department of Health and Social Security, HMSO, London, 1978.

[7] *Short Life Renewal,* Laboratories Investigation Unit, Paper No. 12, Department Education and Science, HMSO, London, 1981.

4

The user's role in the design and furnishing of laboratories

D. L. Morgan
Laboratory Sciences Division, Huntingdon Research Centre

INTRODUCTION

This chapter is directed towards the laboratory user faced with th prospect of moving to new laboratories ans with the decision whether to get involved in their design and if so how to start. It also presents the view that the professions involved in the construction of new buildings need to concentrate more on the purpose and function of what they are designing and less on aesthetics and brilliance of engineering – in short to involve the people who will work in the building.

The author does not seek to present the ideal mechanical and structural design or the perfect internal layout. The aim is to assist the practical involvement of those new to the procedure by guiding them through the stages of the design process with a series of checklists and notes drawn from experience of working in and designing new laboratory buildings.

The user should decide to do two things at the start of a new laboratory project:

(1) Get involved.

If nothing else, just continually asking questions ensures that the designers will have to take account of the user's requirements and problems. There will then be a better chance of getting an environment that works.

(2) Abandon any preconceived ideas about the process and the professionals involved.

The performance and contribution of most of the people concerned will almost certainly be worse than can be imagined. Often the major preoccupation and skill of designers, engineers, manufacturers and builders is in apportioning blame for what goes wrong. There is a pitfall here since without care the user's contribution will be little better. User input is often characterised by frequent changes of mind and a very limited perspective of what the employer requires. There is natural resistance to having an outsider define the working environment. The user should, however, before demanding every service thought to be needed and insisting on rigid designs to accommodate today's equipment and methodology, sit back and think about cost and the advances in technology which could render those ideas obsolete within a few years.

The best approach to designing laboratory facilities is the multi-disciplined project team, with a user's representative as a permanent member. The user will be most concerned with the function of the laboratories and their layout but can and should widen any input to make sure the development reflects accurately the needs of the user's employer.

Table 4.1 illustrates the starting point to this involvement and shows the main stages of the design from the user's point of view. Using this as a framework, each of the key areas will now be considered in more detail.

Table 4.1

Design considerations

Stage	Contributory factors		
Definition of purpose	Organisation's requirements		
Shape/size/position accesses	Organisation's requirements		
	Local planners		
	Costs — projected life		
Overall environment			
Storage/services/laboratories/offices	User's needs today		
Interrelationships			
Degree of flexibility	Organisation's needs tomorrow		
Laboratory planning			
Safety	Fume cupboards, biological		
	cabinets, total enclosures		
	Air supply and extraction		
Services/layout	Degree of integration	Accessibility	
	Fire Officer/Safety Officer	Health & Safety Executive	
	Laboratory function	Storage/workspace relationship	
	Ergonomics		
	Materials		
Environment — walls/floor	Function		
lights/ceilings	Isolation		
colour	Insulation		
	Access		

OVERALL ENVIRONMENT

The overall environment needs to be considered once the shape, size and position of the building has been agreed by the organisation's business planners, architects and local planners. This concerns the inter-relationship between the various functions needed in the building and involves the first critical balance between the user's immediate needs and the employer's objectives, which are probably longer-term. At this stage the services of management services staff and/or architects should be employed to propose alternative overall schemes for building layout which can be reviewed for practicability by the user. It should be remembered that the building will outlast the product line or area of research going into it. It is certain that future alterations will be made even though these cannot be planned at this stage. Planning to make this a straightforward operation introduces the concept of Flexibility.

FLEXIBILITY

It is very important to review Flexibility/Adaptability at the stage when the whole development is under consideration although it is often only considered within each room. In recent years several well-integrated adaptable laboratory building designs have appeared for educational establishments. Little progress has been made in the industrial sector probably because of the different purpose of these laboratories and the failure of those selling the concept to explain the benefits fully to those responsible for running and using them. There is also considerable resistance from the traditional building trades on whose territory the ideas impinge, and from mechanical and electrical engineers who just do not appear to understand them.

Nevertheless some degree of flexibility/adaptability inherent in the laboratory building design is nearly always in the client's interest. Room relationships, servicing networks and operational modules should be designed with this in mind.

Table 4.2 summarises the advantages and disadvantages of the flexible approach and these may be expanded as follows:

Change
    short-term   — workplace configurations/research requirements;
    medium-term   — services changes at point of use.
                       — specialised areas;
    longer-term   — partition walls adaptability.
Obsolescence
    time scale to design and construct;
    legislation changes;
    technology changes.

Table 4.2

The case for flexibility

Advantages
  Allows for change — short, medium- and longer-term
  Helps prevent built-in obsolescence during project development
  Reduces craft input on site — reduces construction time
                              reduces construction cost
                              improves quality control
  Easy decontamination or replacement
  Replace/interchange heavy wear units
  Movable furniture more likely to be depreciated over a sensible life span than if built in

Disadvantages
  Possible higher immediate cost, although not if purchased separately
  Demands good housekeeping
  Often more trouble involved in initial planning and thinking.

Craft input
     furniture and services can be fabricated off-site under controlled conditions;
     site is in the greatest state of turmoil when this is normally done;
     reducing project duration cuts costs.

Contamination
     bench surfaces meticulously sealed to walls and floors are fine until access
     to the services behind them is required, the seal is then broken for ever;
     movable furniture means that everywhere is available for cleaning;
     benches, and even walls, can be disposed of if badly contaminated;
     even with partition walls modern adhesives can ensure a good wall/floor seal
     to contain spillage within operating units.

Depreciation
     built-in furniture is likely to be depreciated over the selected lifespan of the
     building, but lasts only eight to ten years;
     movable furniture can be identified as a separate cost and be depreciated
     over its expected life.

High costs
     often higher construction standards, better materials.

Good housekeeping
     desirable but unpopular, no hidden corners mean no 'out of sight out of
     mind';
     piles of junk mean no flexibility and no easy decontamination;
     nothing permanent should be attached to adaptable walls.

Planning
>tricky interfaces are involved between building fabric and adaptable services/
>walls/furniture;
>architect and furnisher have to earn their money.

SAFETY

The next stage is the actual planning of the laboratory. Safety considerations in
the working environment often impart a degree of rigidity to the building layout.
Current legislation for the protection of staff imposes requirements on designers
and employers and means that the first consideration for most industrial laboratories is the number, position and integration of fume cupboards, safety cabinets
and total enclosures. These items are of primary importance since, because of
the quantities of air consumed and the need for effective performance, they must
be regarded as part of the mechanical design of the building and not treated as a
piece of equipment or furniture to be added at a later stage. Selection should be
guided by: published research from some universities; legislation, for example
for ionising radiation[1]; and relevant Standards[2, 3].

Users should be very wary of mobile recycling fume cupboards unless they
can predict with certainty the type of use and materials worked on. Practical
points to look for are warning lights and easy access to services. The restriction
to the fume cupboard of all dangerous laboratory services, for example gases
under pressure, should be considered seriously.

SERVICES

Services are a problem area and they are often designed, installed and maintained
by specialists who rarely consult with each other. They are expensive, require
maintenance and are a major headache when new layouts, renovations and
redevlopments are carried out.

Under no circumstances should a design engineer be requested to provide
specific items at a particular place, because that is exactly what will happen with
no logic or thought for change, maintenance or visual impact. The design and
installation of services ruins many excellent flexible and adaptable approaches.
It is essential that the project manager and the user review the whole servicing
proposal for its compatibility with the design objectives of the building. A
degree of repetition or better still a grid approach should be looked for and
interfaces with furniture and equipment critically examined. The future need to
transfer data between units should not be overlooked although it may not be
required at the time. The incorporation of cable trunking for local area communications networks is recommended.

LAYOUT AND ERGONOMICS

It is beyond the scope of this chapter to detail the laboratory planning procedure and the ergonomics of good workplace design but the following comments may be of value. It is an area where the laboratory user gains a reputation for being difficult, unprogressive and against change. There is no doubt that the user knows best as far as work and the environment required in the laboratory are concerned. However, rather than insisting on rigid plans and configurations which only reflect local thought, the skills of trained staff should be utilised to solve the problems. Laboratory furnishers, management services staff and architectural technicians under the user's direction will be able to suggest the most efficient and safe layouts to satisfy the objectives.

With a degree of flexibility in the furnishing of the laboratory, the large areas per person referred to in literature on laboratory planning are unnecessary. Scientists are territorial, therefore work space should be fully utilised and large empty circulation spaces avoided. These spaces will not stay empty for long and over-generous allocation results in highly serviced, expensive stores. Studies on storage[4] have indicated the need for an increase in cheap, bulk central stores and less provision for ground-level storage in the laboratory. The basic split between storage in the laboratory and common central storage for the building should be decided when the overall environment is being considered.

Figure 4.1 gives guidance dimensions for classical laboratory work. Providing emergency routes are well delineated, spaces between benches can be limited

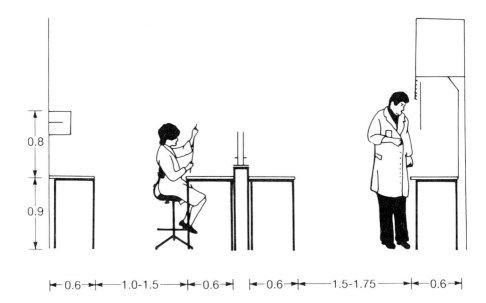

Fig. 4.1 – Layout and ergonomics planning.

depending on staff density. The maximum depth of usable worktop is 600 mm, deeper benches will merely become storage space. Bulky equipment which would overhang would have its own special mounting frame or trolley. The height of work tops is very important and a balance must be achieved to accommodate seated and standing operations. Thus the seating is vital and saving money on poorly designed or rigid seats is false economy. Since scientific experiments tend to be a vertical rather than a linear function, lowering the work top gains volume and in any case it should be limited to a maximum height of 900 mm.

SELECTION OF MATERIALS

The choice of materials used on laboratory benches is usually given little attention and generally it is the manufacturer's main line, inevitably a plastics laminate, which is chosen. There are alternatives to this material as laminates have weak points which must be overcome.

Table 4.3 shows the resistance of basic materials to the laboratory environment and is of the type found in laboratory furnishers' catalogues. The cost factor is not included since costs vary with economies of scale and attempts to promote particular materials. What is certain is that the best all-round materials are expensive and, in addition, tiles, ceramic and stone may give problems at the joints, and stone is heavy. For most situations laminated board offers the optimum balance of durability and economy, but resin glues should be used, and

Table 4.3

Selection of materials for laboratory surfaces

Material	Resistance to							
	Acids	Solvents	Absorption	Abrasion	Impact	Stains	Heat	Water
Stainless steel	2	1	1	1	1	1	1	1
Tiles	1	1	2	1	1[a]	1	1	1
Ceramic sheet	1	1	1	1	1	1	1	1
Slate	1	1	2	1	1	1	1	1
Pierrite (asbestos)	1	2	3	1	1	3	1	2
Formica type	2	1	1	1–2	1[a]	2	2	1[a]
Polypropylene	1	2–3	1	2	1	3	3	1
Hardwood	2	2	2	1	1	2	2	2
Veneer	2	2	2	2	2[a]	2	2	2[a]

Key: 1 good.
    2 adequate.
    3 poor.
[a] Edge detail critical.

edge detail is critical for water and impact problems. A wooden strip stuck on the edge is of little value and post-forming requires careful quality control. Inset hardwood edges can provide a good looking answer to the problem. The introduction of some wood in non-wear areas such as shelves and cupboard sides helps to avoid a sterile appearance.

It is less critical for every unit to be able to resist all types of chemical or impact if unit furniture is used to give flexibility in layout planning. Selected bench units can be ordered made of suitable materials for special tasks and then moved around as required. Sink material is an emotive subject. There is little wrong with the use of polypropylene but in practice it is not kept as clean as sinks made from stainless steel. A fashionable, non-ribbed drainage board will not allow glassware to drain whichever material is used.

LOCAL ENVIRONMENT

Laboratories should be functional and efficient but they also may need to foster creativity and innovation. Laboratory staff will spend most time on furniture selection and layout and overlook the main objective which is to provide a working environment. The factors listed below should be thoroughly developed by the architect. However, without the involvement of the user it is not possible to relate them to the function of the laboratory and the degree of cleanliness and isolation required.

Wall finishes	Treated blocks, plaster or partition finishes in an easily cleaned material?
Paint	Impervious oil-based or water-based?
Floors	Wear, absorption, impact resistance, chemical resistance. Welded sheet without a sponge-back has many advantages.
Ceilings	Suspended (these look better), or direct finish (these avoid dust traps and contamination cavities)? Ceiling finish?
Lighting	Sealed units or standard? Colour. Manual or automatic control?
Interfaces/junction	A very important factor. Walls/window cills/benches. Walls/benches/floors. Is a 'tank' approach needed to limit spills? How are services introduced, are there dust traps? The architect must design the interfaces and not leave the choice to the contractor.
Colours	Rooms are not cleaner because they are white and they are certainly less pleasant to work in.

SELECTION OF THE FURNITURE/LABORATORY SYSTEM

The selection process is where the laboratory user most frequently becomes involved and furniture suppliers play on this aspect and encourage direct contact. Unlike most consultant designers, laboratory furnishers have considerable experience and the astute user can learn a lot about laboratory layout and practical pitfalls from such manufacturers. Nominating at the outset also avoids having the main contractor going for the cheapest joinery quote on the basis of a non-watertight specification.

Table 4.4
Selecting the furniture/system supplier

'Hit list'	
Glossies	
Testing capability: Send out —	specification
	blank plan
	tender invitation
Presentation of proposals	
Evaluation	Is it integrated system?
	Who else uses it?
	Can they design/make furniture as well as sell it?
	Who manufactures it?
	Who installs the furniture?
	Who does the plumbing/electrics?
	Will they accommodate special items?
	Will they agree to include all extras, odds and ends to complete the job?
Test the product	Ergonomics
	Manufacturing standards
	Serviceability

Table 4.4 outlines the main, perhaps fairly obvious, steps in a selection procedure, although it is surprising how many of these the professionals overlook. The following should be read in conjunction with the listed steps:

Use a trade directory for the 'hit list' and include some unknowns amongst those selected. Include a catch, or difficult, area on the blank plans and note who spots it. Make it clear that the suppliers are not being employed as designers but are just being asked to quote.

During the evaluation avoid, 'We make it all to your specification'. It is likely that they do not have a working system, and there is insufficient time or knowledge to re-invent the wheel.

Consider whether the range offers continuity, and whether it will still be made when some additional/replacement units are required. This also applies to architects drawing up specifications and shopping around for each component. The procedure may work for a very large development where spare units are available but it is unwise for most single projects.

The quality control of suppliers who manufacture, fit and pre-plumb at their works is likely to be better than those who carry out the work on-site. Is the interface between mechanical, electrical and building trades a complete, thought-out system which includes services?

All edges should be checked whilst testing the product. The product should be examined by maintenance craftsmen and not by the engineering director. Cabinet strength should be looked for where flexible furniture is chosen.

RECOMMENDATIONS

Any user involved in a laboratory building or renovation project should adopt three points as guiding principles:

(1) Insist on finalising specifications. Provisional Cost sums and contingencies are meat and drink to the construction industry. They are an excuse for the professionals involved to put off difficult parts of the design until the building has advanced to the stage where the best option has probably been eliminated. They inevitably mean extra cost, which benefits everyone involved in the project except the client.

(2) Import as much pre-assembled, pre-plumbed equipment as possible, even the walls. This reduces delays and confusion on site and improves quality control.

(3) Most important of all is never to forget or abandon the original design principles of the project. Constantly remind the project team of what should be achieved, since people on a construction site have ways of overcoming problems which will certainly compromise the principles of the building. A project tends to become an end in itself, being carried along on its own momentum with the original objectives long forgotten.

REFERENCES

[1] *Ionising Radiations (Unsealed Radioactive Substances) Regulations 1968,* Statutory Instrument No. 780: 1968, HMSO, London, 1968.

[2] BS 5726: 1979, *Specification for Microbiological Safety Cabinets,* British Standards Institution, London, 1982.

[3] DD80: 1982, *Laboratory Fume Cupboards,* British Standards Institution, London, 1982.

[4] F. Drake, *Lab. Equip. Digest,* February, 1978.

5

Conflict of safety aspects in laboratory design

N. H. Pearce
University of Bristol

Some years ago the problem of rationalising cost priorities and decision making in new safety projects, particularly in the design of buildings, was examined[1]. Much of the Report is not relevant to modern laboratory design but the early part provides some guidance on tackling real safety issues. Like most theoretical exercises, the Report produced very little in real terms but provided some useful lessons.

In order to establish a mathematical model that could be examined, the problem of smoke travel and means of escape in buildings was investigated. A simple model was constructed of a building designed round a courtyard with doors on opposing sides (Fig. 5.1). The effect of increasing the numbers of smoke doors to the point of unreality was studied and the various effects examined graphically.

The cost of smoke doors in relation to smoke travel was examined. The distance of smoke travel (x), on an arbitrary scale of 1 to 100, was found to decrease with the number of doors (Fig. 5.2). By plotting the reciprocal of the distance of smoke travel $(1/x)$ against the cost of the number of doors, an approximate idea of efficiency in terms of cost for each door can be obtained (Fig. 5.3). It is a well known fact to all concerned with costing that the more doors there are in a building, the less will be the benefit obtained for the cost. In terms of cost efficiency a peak is obtained (Fig. 5.4). Each smoke door has its own inherent accident problem — people stumble, people brush into it, people drop bottles of solvent. In fact, there is a small and normally acceptable risk of accident with the incorporation of each smoke door. This risk is balanced against the efficiency in terms of restricting smoke travel (Fig. 5.5). A similar effect is well known to chemical kineticists in terms of a first order reaction. If the number of doors is increased sufficiently another phenomenon occurs also well known to chemists — the interaction between competitive or alternative reactions.

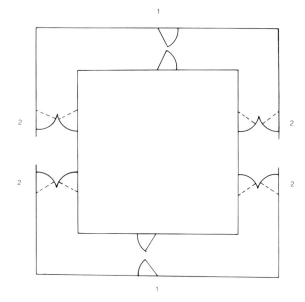

Fig. 5.1 – Plan of test model for investigating smoke travel in and means of escape from buildings.

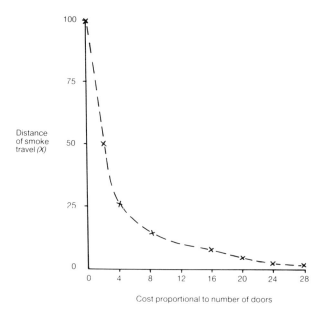

Fig. 5.2 – Relationship of distance of smoke travel to the cost of smoke doors.

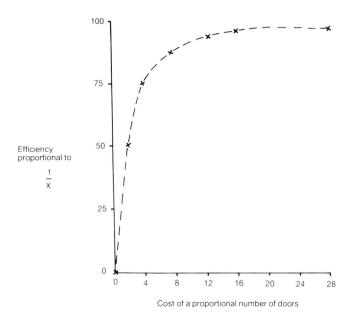

Fig. 5.3 – Relationship of efficiency of smoke clearance to the cost of smoke doors.

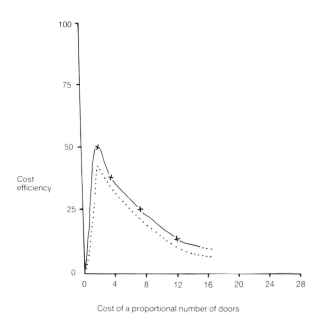

Fig. 5.4 – Cost efficiency in relation to the cost of smoke doors.

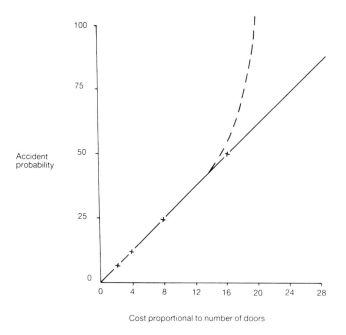

Fig. 5.5 — Accident probability in relation to number of doors provided to assist
smoke clearance.

This occurs if the smoke doors are sufficiently close to one another. One door
interacts with the operation of a second door and increases the accident probability for both. The accident probability at a given point actually increases as
shown by the dotted curve (Fig. 5.5) and is higher than the probability predicted
for the number of doors. This point at which the interactions occur is of great
interest and importance and will be examined later. The conclusion to be drawn
from this work is that because of the interaction effect, the efficiency is by no
means as high as predicted.

This work initiated a more profound examination of the problem of interaction between safety systems. A similar problem is well illustrated by the
counteraction between the efficiency of extracting fumes from a fume cupboard
and conserving the energy within a building. It is also represented by the conflict
between agreement with Home Office licensing requirements to preserve the
lives and safety of animals, and the occasional counter-requirement to maintain a
safe working environment for the personnel. In planning the new investigation it
seemed sensible to continue with this specific problem of smoke doors.

As an example, a corridor with a serving hatch from a chemical store opening
into one side of it (Fig. 5.6) was considered. A typical solution in terms of
fire protection would be to protect such a corridor from a source of fire by

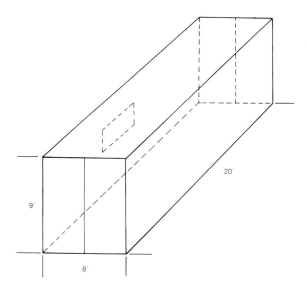

Fig. 5.6 – Diagram of corridor with serving hatch from a chemical store opening
on to it.

constructing smoke doors at the points shown. A very common accident in
laboratory and stores areas is that where a bottle of solvent is dropped before it
is put into the bottle carrier. By estimating the total volume of the corridor
section, and allowing five minutes ventilation at a four volume air change rate,
per hour, which would be reasonable, the effect of such an accident can be
examined. The results from this experiment are shown in Table 5.1.

Table 5.1

Result of benzene spillage in confined corridor space

Total volume of corridor space:	1.440 ft^3 = 40,800 litres
After ventilation (5 min):	54,260 litres
Volume of benzene vapour from 2.5 litre bottle	
(assuming complete volatilisation at 20°C, 760 mm press):	607 litres
Benzene concentration in atmosphere	
No ventilation (smoke doors closed):	$\dfrac{607 \times 100}{40,800}$ = 1.49% v/v
With ventilation (smoke doors open):	$\dfrac{607 \times 100}{54,260}$ = 1.12% v/v

Benzene was used as an example since it illustrates that both toxic and inflammability problems may arise. Since the lower explosion limit of benzene at $20°C$ is 1.40% v/v the installation of smoke doors in this case would achieve a situation in which the possible passage of smoke and possibility of fire has decreased, but a box of explosive mixture has been produced which may be ignited by a spark. The potential energy liberation is of the order of 20,000 kcal. This is more than enough energy to blow out anything other than the load-bearing wall. In view of this and other similar disturbing assessments, further investigation of the problems of interaction has been carried out at Bristol University.

The prevention of fire and the provision of adequate means of escape in the case of fire are matters of general importance. Invariably, this will require the scrutiny of plans on which lines drawn in corridors, etc. represent smoke doors, and dots represent alarm bells, smoke points and hose-reels. No attempt seems to be made to determine whether the smoke detectors are put in places that wind and air actually reaches, or to determine whether bells placed in between smoke doors could actually be heard at the other side of the doors or at the end of buildings. A large number of possible interactions which could cause major problems seem to be overlooked. Despite popular comment to the contrary, none of the difficult, and in some cases dangerous, recommendations seem in the experience of Bristol University's Safety Organisation to originate with the Fire Authorities, which, in fact, have the real statutory responsibilities in this area. Most of the comments of and discussions with the Fire Authorities are very sane, very practical and very realistic. The problems appear to have originated with other bodies and in some cases it has been very difficult to establish who, how or why.

To pursue this problem a little further a number of typical plans were examined and the question asked in each case: What would happen in this particular building if a fire generated a large quantity of smoke? In most cases it was not possible to make any predictions. The essential facts were not presented with or on the plans. How then can the designer or user make predictions? The sensible conclusion is that if it is impossible to know what was going to happen in a major building one should be borrowed to find out. Much modern work is concerned with the adaptation or modification of existing buildings rather than with new. With the very willing assistance of Bristol University's Physics Laboratory staff and the encouragement of the then Director, Professor John Ziman, the Safety Organisation was able to borrow the entire School of Physics for a weekend. Hot smoke was discharged into it (as an old building it did not possess any fire doors) and then simulated fire doors were incorporated in places where it was known they would be suggested (and, indeed, have been suggested) and the exercise repeated. The results were reviewed for each case.

The results of those experiments were both interesting and disconcerting. Figure 5.7 shows a plan of the H. H. Wills Physics Laboratory at the University

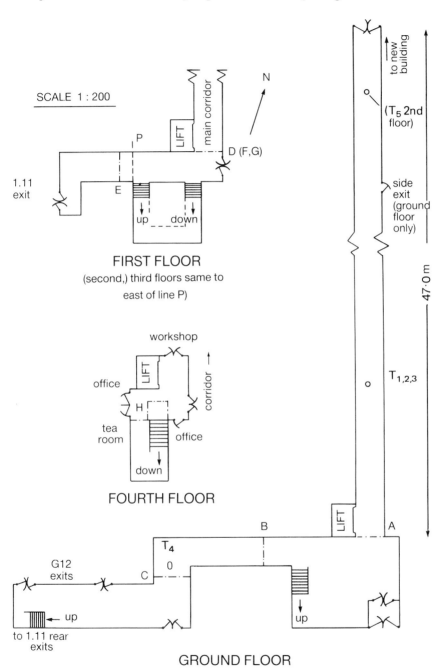

Fig. 5.7 – Plan of corridors in University of Bristol, Physics Building.

of Bristol. This is an extensive building and dates from between the two World Wars. At the front of the building is a high tower in which a very large volume stairwell extends up the four or five floors; in addition there is a basement. The cross-section at the side leads to lecture theatres and the building is linked to a new structure added in the 1960s. One side of the building contains the stores area with access to the basement; a corridor runs lengthways along the building. The building has a new adjunct at the end − a large building enclosing one corridor with offices and laboratories situated on either side of it. There is no interconnection of the rooms on either side and each floor is similar in design to the floor below except for the top floor which has minor modifications. The main corridor is 47 m long. Significant points are the lift in one corner and a large stairwell at the south end with five flights of stairs. Proposals, very common in such buildings, to instal firescreens along the ends of the corridors were examined. The basic idea is to protect the means of escape, i.e. the stairwell, by screening the corridors from the stairwell and hence from fires starting in the main corridor areas and in some cross corridors which lead on to the stairwell. This, in the light of modern safety legislation relating to fire protection, seemed to be sensible.

In the first experiment hot smoke was generated from one of the worst possible situations, namely a point about 13 m along the ground floor corridor. This simulated escape of smoke from a side office or laboratory into the corridor. The first observation was the layering effect of the smoke moving rapidly towards the stairwell. After three minutes the smoke poured into the stairwell. People standing in the corridor would suffer no discomfort. The temperature gradient in the stairwell from top to bottom was 1°C and the degree of turbulence was large. Smoke was transmitted into the stairwell, an action which the smoke doors would be designed to prevent, diluted rapidly, and discharged. The experiments were being carried out in a building in which, since toxic materials are in use, for safety reasons the ventilation cannot be turned off in an emergency. The work showed that, in fact, it did not matter whether, in this building, the ventilation was maintained or not. Various external wind conditions, with doors open and closed and the ventilation systems on and off, were examined, and no major differences in the basic effects were observed. The normal evacuation time for this building is four to six minutes. Photographic records of the experiments showed that the situation had not altered 20 minutes after the initial discharge; smoke was present but it was diluted smoke. There were no problems in passing through it in the stairwell or corridors during the period and it was not possible to block the stairwell. Smoke also tended to move along the cross corridors, become diluted in the stairwell and move up it. Again, no problems of moving along the cross corridors were noticed. After 29 minutes, the smoke pattern retained the layer effect which remained stable despite movements underneath it. The door of the corridor did not become blocked and means of escape to either end, even through the smoke-filled stairwell, were perfectly accessible.

The experiment was repeated with smoke doors located as suggested above. The first effect of the smoke doors was to destroy the layered effect of the moving smoke which piled up in the corridor. In the absence of smoke doors it was possible to evacuate the building for a period of up to 20 minutes. With the introduction of the doors there was no way out after one and a half minutes. The effect of the smoke doors was to remove the safe means of escape. To all intents and purposes most of the main corridor was blocked after two and a half minutes. An additional effect arising from the pressure of the smoke doors was that smoke was being forced into ducts and, in particular, into the lift shaft which previously had been unaffected. Thus, as well as the smoke doors actually blocking the corridors in which the smoke was generated, smoke blockage occurred in corridors and some cross corridors on upper floors via the lift shaft and ducts. This situation had not been observed in the first experiments. Within five to six minutes of the liberation of smoke, three corridors were effectively sealed. The means of escape at the first floor level was blocked by smoke passing via the ducting and the cross corridor became impassable. After nine minutes, the situation became intolerable. Rolling back the simulated smoke doors at the southern end of the corridor caused the corridor to become impenetrable. Escape from offices along the corridors was now impossible.

Table 5.2

Dimensions and conditions prevailing during experiments carried out in the Physics Laboratory, University of Bristol

Volumes of corridors		
Ground Floor	380 m^3	
1st Floor	380 m^3	
2nd Floor	360 m^3	
3rd Floor	400 m^3	
Cross corridor at south end of Ground Floor	340 m^3	
Cross corridor at south end of 1st Floor	110 m^3	
Volume of south stairwell	1310 m^3	
Average air temperature	$21°C$	
Temperature at top of stairwell	$22.5°C$ (approx.)	

External winds: Strong; gusting to 10 m/s in the morning, abating in the afternoon.
Steady linear velocity <1 m/s in approx. SW direction at about $30°$ to the south face of the building.
Internal air currents <0.1 m/s, with no detectable variation with external wind or with the building fans on or off. Door and windows at south end of building closed.

A similar situation was investigated in the cross corridors across the face of the building at the ground level between the lecture theatres and the stairwell. Hot smoke was liberated at a fairly high rate in these corridors. In the absence of smoke doors, some concentration of smoke was achieved but because of the sink effect of the stairwell it was never possible within a reasonable time period actually to block the cross corridors to the point at which they became unusable. Thus a major safety criterion, that of maintaining a means of escape, was met.

In another experiment, smoke was liberated on the second floor of the building from a point adjacent to a chemical laboratory. A 'backing' effect due to the smoke door at the north end of the corridor adjacent to the new wing was observed within one minute of liberating the smoke. The smoke was so impenetrable that it was hardly possible to see the walls on either side. The absence of a smoke door at the end of the corridor opening on to the stairwell assisted the clearing effect of the stairwell.

Some of the important dimensions and conditions prevailing during the experiments are shown in Table 5.2. The work described in this chapter has been repeated in other buildings and is continuing. The assistance of Dr M. G. Priestley (Dept. of Physics) and Mr A. J. B. Cruikshank (Dept. of Physical Chemistry) with this work is gratefully acknowledged.

CONCLUSIONS

In the particular building investigated there was no way that the type of prevention advocated and relating to the *Offices, Shops and Railway Premises Act 1963* could be accepted. This Act, for some mysterious reason, appears to be extrapolated as the guide for the design of laboratories.

The major conclusion reached so far is that each building behaves differently and must be considered in its own right. Results obtained in the Physics Building were different from those obtained in a building used by the School of Chemistry, but in neither did existing codes of practice help in predicting the correct siting of smoke doors, etc. In fact, smoke came out of all sorts of places which were not predictable from an examination of the plans. The practice of drawing lines on plans in the absence of detailed information causes concern for those responsible for safety.

It is clear that a major factor when designing means of escape is not distance of travel for personnel but the ventilation rate and the direction of air movement within the building. The occupancy level surrounding the areas of hazard and the adjacent work occupations are rarely considered but are a matter of prime importance. Questions very rarely asked by many of the enforcement agencies concern the interactions within legislation. The experiments with smoke doors illustrates only one of the aspects concerned in design but demonstates this point well. In some cases existing legislation actually counteracts a safe design.

The views of local professionals familiar with local conditions and local problems should be sought early in the design stage if safe laboratory accommodation is to be achieved.

REFERENCE

[1] N. H. Pearce, The application of cost benefit systems to safety projects, *Proc. Annual Conference University Safety Officers Association, Bristol 1979.*

6

New developments in laboratory design in the United States of America

J. R. Moody

National Bureau of Standards, Washington DC, USA

INTRODUCTION

This chapter is based on experience gained at the United States of America (USA) National Bureau of Standards (NBS), although many of the practices to be described are not unique to that organisation. Because of the author's background, many of the examples given are specific for analytical laboratories. Examples of current trends at NBS as well as in laboratories which have won the *Industrial Research and Development* Laboratory of the Year Award are included.

The NBS laboratories were built during the early to mid-1960s and more than a dozen major laboratories accommodating more than 2,500 employees are distributed over about 0.4 hectares. Since the building designs took into account the cost of energy in the 1950s and the 1960s, the present cost of energy makes energy conservation an issue at the NBS laboratories. Many of the conservation efforts which have been undertaken at NBS have been duplicated by other laboratories. The task force management system devised to study energy conservation at NBS has been described[1].

TRENDS IN ENERGY CONSERVATION

A review of recent history since the first oil embargo in 1973/74 is out of place in this chapter but the information in Fig. 6.1 is instructive. The graphs show the unit costs in dollars per million British thermal units (Btu) for the three basic energy utilities. With the exception of a few electric generating plants, nearly all utilities in the USA are privately owned and receive no public subsidy. Both oil and gas costs are still considerably less than in Europe since the USA has considerable reserves of both, and the cost of domestic oil and gas is still well below current world market prices.

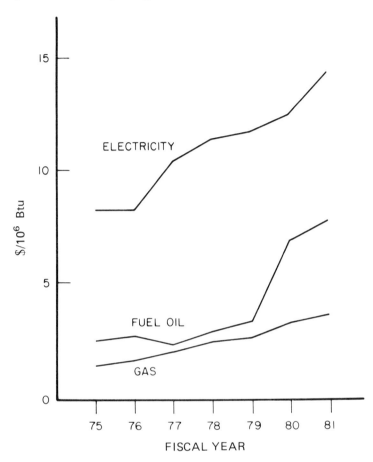

Fig. 6.1 – Average unit cost of energy in the USA.

Presumably because of the current economic recession, the world price of oil has been relatively stable, and, in the USA the cost of oil has not risen above the levels indicated in Fig. 6.1. However, the gap between oil and natural gas continues to decline as the price controls on USA domestic natural gas are phased out. Electricity utility costs are likely to continue to grow at a similar rate in the future although the current halt in fossil fuel price rises may delay future price increases of electricity utilities. Nevertheless, what all managers have to face is the upward spiral of total costs for energy such as those experienced by NBS (Fig. 6.2). It should be said that these figures represent costs incurred at NBS despite a relatively successful energy conservation programme (Fig. 6.3).

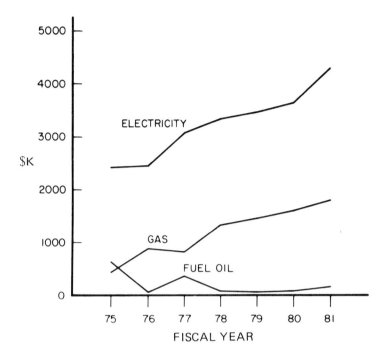

Fig. 6.2 – NBS expenditure on energy.

Clearly, high energy costs are now a significant factor in the operating budget of all institutions. All laboratories consume energy and many of these energy costs, such as for instruments, air flow, and humidity control, cannot be diminished easily without impairing the mission of the institution. Older buildings tend to have more energy problems than new buildings, because of their design. Climate is also a factor. The NBS laboratories in Gaithersburg have both a significant heating requirement in winter and a significant cooling requirement in summer. The climate is, for example, much less temperate than in London.

In November 1973, the then Director of NBS, Dr Richard W. Roberts, appointed Dr John D. Hoffman, Director of the Institute for Materials Research, to head a Bureau-wide task force on energy conservation. The task force was made up of scientists, managers, facility engineers, and staff. The recommendations of the task force were published in 1974[2]. At that time, annual energy consumption was about 115 million kWh of electricity and about 730×10^9 Btu of heating fuel. Of the total energy used, approximately 85 per cent was for

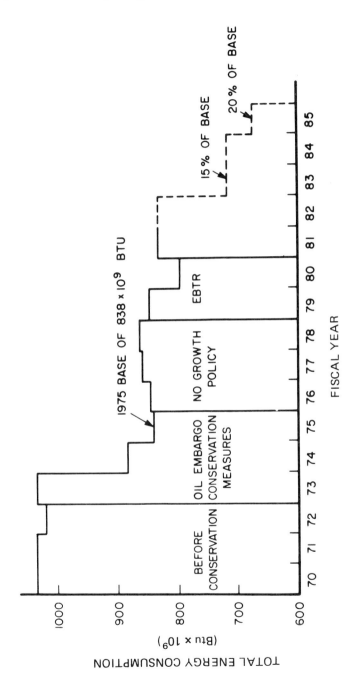

Fig. 6.3 — Total energy consumption, NBS, Gaithersburg.

climate control and the remainder for equipment and lighting. The scientific staff took part in efforts to determine what the actual needs of the institution were, and a complete systems analysis was made of how energy was consumed. Two broad sets of recommendations were made which may be categorised as immediate conservation steps that required little cost or action to implement, and longer range conservation steps which would require significant time and/or expenditure.

The low cost measures were implemented as soon as possible and the results of these measures are represented in Fig. 6.3 for the fiscal years 1973 to 1975. In 1974 the conservation measures produced savings of about 12 per cent in electricity and 18 per cent in heating fuel. These measures included lighting reductions, thermostat adjustments combined with selective building and zone shutdowns, and changes in cooling coil parameters. One unexpected result of the analysis of the NBS laboratories was the finding that operating electrical power could be reduced by 40 to 50 per cent for a few hours on a contingency basis without terminating long-duration and key experiments or shutting down critical facilities. Even without the need for conservation, there was a need for such a laboratory disaster analysis. Mathematical models were developed to represent the climate control system which in turn was useful in evaluating conservation options as well as the impact of changes in climate control parameters on the laboratory environment.

During several years following, numerous more complicated approaches to energy conservation were completed. These included the eventual replacement of all ballast transformers in the hallway lighting fixtures with new ballast transformers designed to handle two fluorescent bulbs instead of four. The saving in electrical energy was significant while the slightly reduced lighting levels in the hallways was not considered objectionable. Other examples included the overnight and weekend shutdown of air handling units in office modules and the selective shutdown of some fume cupboards during the same time period. The years 1976 to 1978 saw a slight growth in fuel consumption (Fig. 6.3) during the no-growth policy period but this was halted during 1979/80 as the Emergency Building Temperature Restriction policy was announced by President Carter.

The historical trend at NBS has been that of gradually increasing energy consumption. If the conservation progress made during 1973/74 is eliminated, then the total energy consumption of the institution from 1975 to 1981 grew by only 0.55 per cent. This is an indication of significant progress in energy conservation, since no laboratory project has been affected in any way by the conservation programme.

Finally, during 1981/82, major studies of the buildings were completed. The most obvious retrofit projects in terms of window sealing, wall insulation, air handling control testing and modification, and corridor lighting have been completed. A number of retrofit projects have been identified as a result of the building energy audits and a study of the steam and chilled water systems. The immediate goal of these projects is the attainment of a 20 per cent reduction in energy consumption.

Retrofit possibilities in the steam generation plant include waste heat utilisation and emphasis on better control of oxygen trim. These combined with other control modifications in the air handling units would achieve 41 per cent of the energy reduction goal. The estimated cost of these retrofits is $435,000 which should be recovered in three years. A considerably more expensive option is the conversion of the existing steam plant to coal gasification. Solar energy options have been found to be practicable if they are limited to pre-heating water for the existing domestic hot water system.

Obviously, for such a large site, it is not practicable to outline fully the specific goals and accomplishments of the energy conservation programme at NBS. What is significant is that funds in excess of $1 million have been made available for these projects and very substantial savings in energy costs have already been realised. The growth in the energy consumption curve has been halted, even slightly reduced, over the past decade and there are good prospects that in the near future substantial improvements will be made as more retrofits are completed. The importance of these projects should be more obvious if energy costs continue to rise, and it is reasonable for laboratory planners to expect these rises to occur.

What has been accomplished at NBS has been similarly achieved in many, if not most, Federal laboratories and institutions over the past decade. Many consulting firms have gone into business to exploit the market for energy audits and retrofit engineering. It would be difficult to assess the degree to which private industrial laboratories have instituted energy conservation programmes; however, it is worthwhile to note that many large chemical companies instituted radical energy conservation programmes to control their product costs, especially on petrochemical feedstocks. Some of these programmes have been so successful that the companies now market their energy conservation skills just as they have already marketed their analytical services.

Thus, one obvious trend in laboaratory design in the USA is towards energy conservation whenever building renovation or modification takes place. With new building construction, an opportunity exists to take advantage of modern, passive solar architecture and indeed many laboratories are taking this opportunity. For example, the 1980 to 1982 competitions for the *Industrial Research and Development* Laboratory of the Year Award have featured many laboratories utilising abundant natural lighting through the use of clerestory roofs, atriums, and other architectural features designed to capture, direct, and diffuse natural lighting into the laboratory or office[3].

Some of these laboratories include the Sherex Chemical Co. laboratory near Colombus, Ohio (winner of the 1982 award); the Arco Laboratory, Newton Square, Pennsylvannia; the C-I-L Inc. Chemicals Research Laboratory, Missanuga Ontario, Canada; and a new Phillip Morris laboratory in Richmond, Virginia, which was not completed during the selection period for the 1982 award. Previous award winners including the Halliburton Services Research Center, Duncan, Oklahoma (1981), and the Alcan Kingston R and D Centre, Kingston, Ontario, Canada (1980), also made notable use of natural lighting.

Other than the obvious savings in illumination costs, the judges noted that numerous other advantages made these features desirable and helped to make the laboratory a desirable place in which to work. All the laboratories mentioned are quite striking in appearance and demonstrate a high degree of good integration with the site and a concern for the interaction of people with their work environment. The sponsors noted that considerable effort had gone into the presentation of laboratory designs for the competition and that energy conservation was a common concern[3]. Table 6.1 lists all the Laboratory of the Year Awards to date.

Table 6.1

Industrial Research and Development
Laboratory of the Year

1982	Sherex Chemical Co.	Dublin, OH
1981	Halliburton Services Research Center	Duncan, OK
1980	Alcan Kingston R & D Centre	Kingston, Ont., Canada
1979	No award presented	
1978	Marjorie B. Kovler Viral Oncology Laboratories	Univ. of Chicago, IL
1977	Ivan A. Gelting Laboratory Aerospace Corp.	Los Angeles, CA
1976	Arapahoe Chemical Inc.	Boulder, CO
1975	Lawrence Livermore Laser Fusion	Livermore, CA
1974	Philip Morris Research Center	Richmond, VA
1973	Squibb Institute of Medical Research	Princeton, NJ
1972	NRC Microelectronics Research and Manufacturing Laboratory	Miamisburg, OH
1971	University of Cincinnati Graduate Engineering Research Facility	Cincinnati, OH
1970	Kaiser Aluminum and Chemical Corp. Center for Technology	Pleasanton, CA
1969	Duquesne University Mellon Hall of Science	Pittsburgh, PA
1968	Batelle-Northwest (Richland Research Complex)	Richland, WA
1967	Bell Telephone Laboratories	Holmdel, NJ

WORK ENVIRONMENT

Fortunately for the laboratory occupant, it would appear that architects now devote considerable attention to the work environment[4]. As previously noted, energy efficient designs are frequently capable of creating very desirable and efficient laboratory and office space. If there is a discernible trend it would seem to be in the use of a campus style of landscaping. This seems to be a common feature among laboratories in the *Industrial Research and Development* competitions. By incorporating nature into the design of the laboratory site, a very pleasant work environment is created. The NBS laboratories, for example, have very large grounds consisting of wooded areas, open areas and small lakes which are a source of pleasure for many of the staff through all seasons. Inside the laboratory building, the availability of office space apart from laboratory space is a desirable design feature and virtually a necessity for professional staff. Other facilities such as meeting rooms, libraries, cafeterias, and other specialised areas are all desirable in fostering an efficient and humanistic approach to laboratory design. A somewhat dated National Institutes of Health standard set a minimum space for research related work of 65 per cent of the net available area. Today that figure is low, especially for private laboratories[4]. Thus, even with all the planning for employee facilities and comforts, no real loss in the efficiency of use of available space is necessary.

New laboratories in the USA now provide special access and facilities for the physically handicapped. In many cases this need not raise costs; traditional entry doors, for example, are a barrier to the handicapped and often an impediment to the elderly or weak. Properly designed entry ways need not have stairs or entrance doors which impede either the physically able or handicapped. Equal access is now required by law in the USA and many older laboratories, NBS among them, have provided retrofits which are often quite expensive. Nevertheless, the 1982 Laboratory of the Year entry cost about $100/ft^2 or about £500/m^2 exclusive of furnishings[2].

LABORATORY SAFETY

It is not likely that anyone has ever deliberately designed a facility to be unsafe. All those concerned with laboratory design basically desire to do things in a safe way. Nevertheless, it is hard to escape the conclusion that the climate for safety consciousness has never been greater. At least part of this is a natural result of the increasing knowledge about industrial hygiene and health. Indeed, it is only necessary to read newspapers to see evidence of the public's concern for such matters as nuclear waste and carcinogens.

Quite often this concern has been translated into legislative action which in turn now affects laboratory operations. In the USA, the Occupational Health and Safety Act has led to a number of government regulations which control permissible exposures of workers to toxic materials. Twenty years ago, a safe laboratory would have been one which depended solely upon the professional skill and knowledge of the individual scientist to avoid trouble. This approach undoubtedly led to many exposures of the then unknown hazards which would be regarded as intolerable in the light of current knowledge.

For most laboratory designers and users, government regulations and exposure limits are facts of life. The exact consequences of these regulations will depend greatly upon the nature of the laboratory work, but it seems safe to say that chemistry laboratories are affected more than others. The hardware required in the laboratory may seem simple but considerable thought should be given to its location and use. Typical equipment found within chemistry laboratories includes safety showers, eye washes, fire blankets, fire extinguishers, and fume cupboards for the extraction of noxious vapours. Of all these, adequate fume exhaust seems to be one of the most effective ways to keep exposure to toxic materials within acceptable limits.

However, there are many more hazards to be considered, especially in a modern large analytical laboratory. Lasers, pathogens, carcinogens, acids, and many other classes of hazard require specific measures for protection of the scientist. At NBS, this has led to the design of special laboratories or facilities which are definitely not flexible in use or design principles. This trend is not limited to NBS and in the author's experience, the tendency to build these specialised facilities is greater in the USA than in Europe.

Other examples will be given later, but one example of a special-purpose laboratory designed specifically to solve a safety problem is the new NBS Toxic Materials Handling Laboratory. This facility permits the safe handling of carcinogens and other toxic materials during those operations necessary to prepare the samples for analysis. Both high efficiency particulate filters and activated charcoal filters are utilised in the design of the laboratory. The facility was designed in-house, and required several man-years effort and more than $100,000 to build.

Certainly the facility was designed and built in response to a perceived need and it would appear that the recognition of the potential of a hazard is the same on both sides of the Atlantic. However, to judge from an admittedly small group of Europeans who have seen the plans, it would seem that the American response to the problem was perhaps more elaborate than similar reactions in Europe. The intention has been to provide a laboratory environment as safe as deemed necessary or believed to be necessary at a particular time. Where standards are lacking, the intent is to build with safety standards that will take into account any conceivable problem.

TRENDS IN LABORATORY ORGANISATION AND THEIR IMPACT ON LABORATORY DESIGN

Ten or twenty years ago, the chances were that laboratories would have been neatly organised in terms of function. One laboratory might have been used solely for optical microscopy, another for spectroscopy, another for measurement of dielectric constants. Since then laboratories have become much less neatly compartmentalised, and interdisciplinary science has had a much greater influence. The reasons for these changes are numerous and probably well known. One consequence has been the complete reorganisation of entire laboratory complexes[5].

As a result, it is very likely that the present-day visitor to a laboratory is likely to see space organised by task, such as trace organic analysis. Because these changes are not likely to be permanent, the issue of laboratory flexibility is real and more important today than ever. Some measures used to achieve this flexibility are discussed later. Rather than focus on the specific changes in the organisation of many laboratories, an illustrative example will be made of the NBS analytical laboratories to show how their design has changed in response to institutional reorganisation.

When the laboratory was relocated to its present site in the late 1960s the building designs were the most modern and flexible available at that time. The entire laboratory complex is built around a standard laboratory module. Modules may be combined in a variety of ways and may be fitted out with any standard laboratory service. Provisions were even made for the supply of services in the future that could not be predicted based on present need. Generally, each module houses a specific instrument or process and usually two scientists.

The organisation of the analytical laboratories was broken down into ten disciplinary units or sections such as electrochemistry, mass spectrometry, activation analysis. Each section had a number of laboratories housing each of its constituent specialities. Today, a larger group of analytical chemists is organised into just three divisions – Organic Analysis, Inorganic Analysis, and Gas and Particulate Analysis. Numerous changes have occurred in laboratory design, but not necessarily all for the same reasons.

One evidence of change is the number of shared facilities, special-purpose laboratories, and large laboratory complexes that were unknown 10 to 12 years ago. Management by task or mission rather than by scientific discipline has led to the creation of facilities which were beyond the economic means of any single scientific division. Certainly it seemed better to have part-time use of a shared facility rather than nothing at all. A good example of this is the trace element, clean laboratories[6]. Few institutions could afford to provide separate clean laboratory facilities for every discipline or scientist desiring one. Such facilities can be shared and in time others may be provided. The NBS Center for Analytical Chemistry now has three clean laboratories with other clean facilities being planned for the future.

Other specialised laboratories built recently include a new Purified Reagents Laboratory[7] and the previously mentioned Toxic Materials Handling Laboratory. This trend towards specialisation of the laboratory seems to be growing, and of course it is contrary to the concept of flexible laboratory planning. These facilities are expensive to provide and, because of their uniqueness, they tend to become permanent fixtures until they are scrapped. The justification for such facilities is based entirely upon scientific need and the willingness to pay the cost to obtain the desired laboratory performance.

Another obvious trend is shared special facilities which reduce overall cost and/or provide facilities which might not otherwise be affordable. Certain high cost analytical instruments also lend themselves to sharing by different groups of researchers. Examples of such instrumentation include Fourier transform infrared spectrometers and nuclear magnetic resonance instruments. Some of these instruments are physically so large, or have special requirements so unique, for example, very accurate temperature regulation, that the entire laboratory must be specially designed to accommodate the instrument. This makes the laboratory somewhat inflexible, but the concept of laboratory flexibility is useful only so long as it serves laboratory needs. When available flexible laboratory designs are technically inadequate, then the only alternative is to build a dedicated laboratory.

Shared or clustered laboratory facilities have also proved to be advantageous when a number of similar instruments are used. For example, gas chromatography and liquid chromatography instruments can be clustered effectively since they can share many of the same peripherals. Similar arrangements have been employed for plasma and emission spectrometry, and other spectroscopic laboratories. Almost all the laboratory remodelling performed in the last few years has been multiple module in size. This trend towards larger, more open laboratories is not unique to NBS and it may be just the most visible effect of the shift away from the one-person laboratory [5].

FLEXIBLE LABORATORY DESIGN

Raab [5] attributes laboratory design in the past to the influence of the academic laboratory which was used as the model for industrial laboratories. In the past, research tended to be on an individual basis, but today, according to Raab, a large number of reasons, primarily regulations, have caused fundamental changes both in the way researchers work as well as the structures they work in. Flexibility in laboratory design has become an important issue as industrial laboratories must constantly change to meet new requirements. In the author's opinion, based solely upon personal travel to laboratories in the United Kingdom and Europe, American laboratories and equipment manufacturers are more conservative and many of the fittings and hardware used in flexible laboratory design are not readily available in the USA.

The NBS laboratories were designed by a committee which included scientists, engineers, and architects. After some 17 years of service, the laboratories continue to show evidence of fundamentally good design. All laboratories are on the inside of the building, for temperature control, and offices are on the outside. Each general-purpose laboratory building has several hundred office and laboratory modules each approximately 25 m^2 in area. The wall panels are of steel unitised construction and the partitions are easily moved or removed to suit laboratory needs. Each module has independent air conditioning/heating control and supply, and independent fume exhaust.

The common inside corner of four adjacent laboratories shares a vertical chaseway which contains the exhaust ducts for the individual modules of either three or four floors of laboratories along the chaseway. The upper floor contains all the building's mechanical equipment. Sandwiched between the entry doors to the modules are similar chaseways for all of the laboratory services. These services are brought into the laboratory along the partition walls between laboratories. In the absence of such a partition, they may be brought in through a trench built into the floor. Thus, any laboratory may receive or reject any desired service completely independently of any other laboratory. Services are, however, permanently plumbed in and are therefore not flexible in the most modern sense.

The high rise research tower at the Philip Morris laboratory contains similar modular design of laboratories on each floor. The provision of services to these laboratories was unique in approach and undoubtedly was one of the factors leading to the Laboratory of the Year Award in 1974. A striking feature of the external architecture is a series of brick finished ribs or fins which rise from the first floor to the roof. These contain the ductwork from the laboratories and neatly solve the problem of proper discharge of fumes by doing so at high elevation and well removed from the air intake for the laboratory.

Laboratory services are brought through the floor and may be reached by lifting the false ceiling in the laboratory on the floor below. A service access corridor built into the outside wall of the building provides access to the main vertical risers of the laboratory services. Both the NBS and Phillip Morris tower represent good examples of modular laboratory design. Some laboratories have gone completely towards the open laboratory concept. One example of a laboratory with considerable open laboratory space is the US Geological Survey Water Resource Quality Laboratory in Arvada, Colorado.

Raab[5] credits L. I. Kahn with introducing the concept of interstitial space. This concept was utilised at the Jonas I. Salk Institute in La Jolla, California, and features independent, isolated, full height levels sandwiched between floors to provide laboratory services to the laboratory spaces above and below. The stated advantages include complete freedom for placement of laboratories, offices, corridors, etc. anywhere on any floor. In addition, all of these may be changed around, thereby permitting the addition, removal, or adjustment of services without interrupting work in progress in adjacent space. This concept of complete freedom in the arrangement and rearrangment of laboratory space would appear to be the direction in which current laboratory design is moving. It would also appear that many of these open, flexible designs are also quite compatible with the solar architecture previously mentioned.

CONCLUSIONS

A number of factors, including energy, safety, government regulations, budget restraints, and changes in research instrumentation, have led to profound changes both in the structure and operation of laboratories. At NBS these changes led to the construction of large, dedicated but less flexible laboratory complexes which were required for the accomplishment of particular tasks. There has also been a shift away from individual laboratories toward larger, more modern, and even shared facilities. Energy conservation and safety concerns have also had an influence. Elsewhere, in newer laboratories, the influence of energy efficient or solar design is evident together with a trend towards open or more flexible laboratory design.

REFERENCES

[1] R. W. Roberts and J. D. Hoffman, *Res. Mgmt.,* 1975, **18**, 26.
[2] J. D. Hoffman. *Energy Conservation at the NBS Laboratories,* NBS IR 74–539, US Government Printing Office, 1974.
[3] H. L. Thomas, *Ind. Res. Dev.,* 1982, **24**, 102.
[4] F. L. Bernheim and J. Sondin, *Ind. Res. Dev.,* 1982, **24**, 112.
[5] M. D. Raab, *Ind. Chem. News,* 1982, **3**, 24.
[6] J. R. Moody, *Anal. Chem.,* 1982, **54**, 1358A.
[7] J. R. Moody and E. S. Beay, *Talanta,* 1982, **29**, 1003.

7

Designing laboratories for a tropical climate

M. Danby
School of Architecture, University of Newcastle upon Tyne

INTRODUCTION

It is not possible in one chapter to present a comprehensive guide to the design of laboratories in tropical climates. The emphasis, therefore, has been concentrated on the differences between the situation in a technologically advanced country and that which prevails in a tropical country which is almost certain to be part of what is presently known as the Third World.

Before defining the differences it is necessary to ascertain the similarities, because the majority of publications concerning laboratory design are written by those working in technologically advanced countries and assume similar conditions. Pure science and the image of the scientist are universal; applied science is almost so. Laboratory equipment, space requirements and safety standards are practically the same in principle throughout the world, although some countries may have a temporary lead in the design and supply of sophisticated apparatus. Most Third World countries, however, must import laboratory equipment from technologically advanced countries. Therefore, the detailed technical requirements, including space and safety standards, will not be appreciably different in a tropical country although the means to achieve and maintain them may well be so (Fig. 7.1).

Similarly, the requirements of the scientist and ancillary staff in relation to lighting will be the same anywhere in the world. Human needs in relation to thermal comfort are universal, with some variation due to acclimatisation and type of clothing, but again the means to achieve them will vary because of differing external climatic conditions and the difference in available human and economic resources. In order to obtain satisfactory thermal comfort contitions within a laboratory it is essential to understand how the building fabric interacts with the external climate and to do this, certain basic climatic data must be available to the designer.

Fig. 7.1 — Interior of Computer Centre, University of Khartoum. International in character without local connotations and constructed within an existing brick building at least thirty years old.

CLIMATE

The first basic difference is that of climate. Tropical climates vary between the two extremes of hot/dry and hot/humid. Hot/dry is, as the name implies, both predominantly hot and dry throughout the year, but this does not mean that there is not heavy rainfall at some time or that the humidity cannot rise when there is little or no rainfall recorded. Coastal situations can be very confusing, and places like Karachi and Port Sudan can experience relatively high humidity over several months of the year when the prevailing wind is passing over the sea. This latter type of climate is often referred to as hot, dry maritime. An extreme hot/dry climate with little or no rainfall and no humidifying winds will usually result in desert conditions with little static human settlement unless there is a river to bring sufficient water to make cultivation and permanent human settlement possible. The Nile is an example of such a river. The other extreme of the tropical climate scale is the hot/humid climate which is obviously predominantly hot and humid throughout the year. The rainfall is consistently high, normally producing dense forest conditions.

Between these two extremes are a range of climatic possibilities with wet and/or dry seasons of varying length. It is common in some parts of Africa to have two wet seasons, one dry season and one, relatively cool, moderate period. Those parts of the world affected by the monsoons can experience varying extreme climatic conditions. Elevation above sea-level can bring yet another moderating effect of cooling. Mountain areas in the tropics can experience cool even cold conditions at certain times of the year.

From this brief sketch of climatic characteristics to be found in the tropics, it follows that it is necessary to have monthly average data for a full year to understand fully the climate and the seasonal changes. Air temperature, rainfall, relative humidity, sunshine hours and direction and force of prevailing wind presented in graphic form (Fig. 7.2) can be used to indicate to the designer the external forces that the fabric of his laboratory building must modify to obtain the desired internal environment.

DESIGNING THE LABORATORY ENVIRONMENT TO MINIMISE THE EFFECT OF CLIMATE

External factors on which human response to the thermal environment depend are air temperature, humidity, air movement and radiation. There are, however, some individual factors such as clothing, activity, degree of acclimatisation, age and sex. To assess the internal environment of a laboratory which will produce human comfort conditions is therefore fairly complicated. Experimental apparatus and materials may also have their own narrow requirements of temperature and humidity. It is, therefore, only possible to aim for a range of temperature and relative humidity which will satisfy all the required conditions. This range can be established as the desired comfort zone and compared with data of the external climate to discover when overheating or underheating occurs [1]. The desired role of the building fabric in the modification of the external climate can then be assessed. Although overheating is obviously a more common occurrence in the tropics than underheating, it is often the case that any necessary cooling is

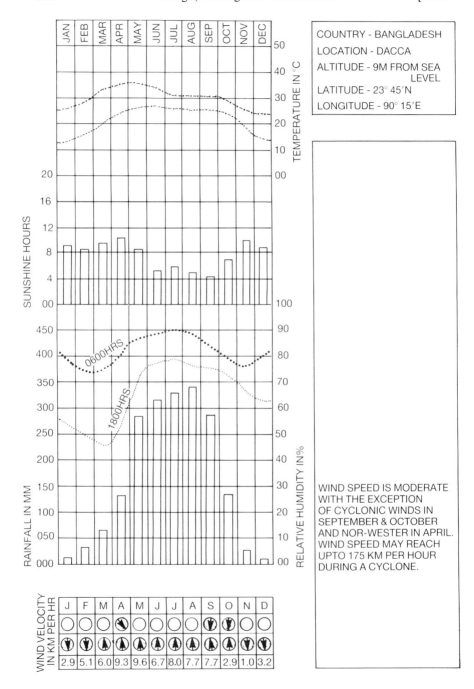

Fig. 7.2 – Climatic data for Dacca, Bangladesh [2].

achieved by mechanical means rather than relying on skilled design to enable the building fabric itself to produce the desired internal environment by passive means. Although there exist many examples of local traditional architecture in tropical countries which do just this, there is still the tendency, particularly in prestige projects, to ignore the possibility of passive means and rely totally on air conditioning which is high both in capital and maintenance costs.

There is, however, one passive device that has fairly universal acceptance, and that is the 'sunbreaker' which prevents the ingress, through windows or other openings, of direct radiation to the interior of a building. The design of the sunbreakers should present no difficulties since the sun angle can be predicted for any time in the year and for any orientation of the building. As they are exposed to solar radiation, they should be reflective and of the minimum thermal capacity possible. Shading devices should always be used in preference to anti-sun glass which is expensive and often difficult to replace when damaged.

Obviously, there are occasions when passive designs cannot alone create the required comfort conditions especially when heat is produced internally by humans, in the case of teaching laboratories, or by scientific experiments and processes themselves. In hot/humid climates, air movement is desirable in order to cool the damp skin surface by evaporation and the more humid the air the greater the volume of air needed to achieve evaporation. Natural through-ventilation can be obtained by correct planning and design of openings in external walls and is often supplemented by the use of ceiling- or wall-type electric fans.

In hot/dry climates external warm air can be blown by an electric fan through a filter made from wood fibre which is kept constantly moist by a water spray. This air is cooled by evaporation before it passes to the interior and the apparatus is known as an air cooler. This principle is the same as that used by many traditional devices in the Middle East, using wind pressure in place of electric fans. It is cheaper in capital and running costs than full air conditioning which normally requires the use of compressors for its cooling effect. The air cooler cannot, of course, be effective if the external air has already a high relative humidity, but it does have the possibility of switching off the water spray during the rainy season using only the fan for forced ventilation. This is useful in those hot, dry areas which experience relatively short seasons of rain and humidity. If full air conditioning is needed there are the small unit types which can be fixed in an external wall and are suitable for relatively small spaces such as preparation rooms, stores, offices and research rooms. If the space to be air conditioned is larger and especially if there are many such spaces, then a central plant is required with distribution ducts for cool air. Alternatively, chilled water may be distributed through insulated pipes with individual air blower units where needed. The latter system is usually more economic.

Dust may also be a problem in hot, dry areas. It can damage equipment, create a health hazard and necessitate continuous need to clean laboratories.

Efficient sealing of doors and windows and the creation of a pressure differential can help to minimise this problem.

THE INFLUENCE OF INDIGENOUS RESOURCES AND SKILLS IN LABORATORY DESIGN

A sensitive understanding of climate is an essential requisite for the designer of science buildings in tropical countries, not only for the reasons already given but also to reduce the demand on resources and skills.

The majority of tropical countries suffer from a shortage of resources (Table 7.1). There is usually a scarcity of capital but many scientific projects are treated as a matter of prestige, so in some cases capital is sometimes made available locally or financial aid is obtained from outside sources. Even in these cases it is rare that recurrent money is made available on a regular basis for subsequent care and maintenance as well as for the replacement of parts and equipment when needed. These difficulties are accentuated if the initial design assumes an unrealistic level of supporting infrastructure of services which are often unreliable, and skills which are particularly scarce at technician level. For reasons too complicated to discuss in this chapter, technical education has so far failed

Table 7.1

Selection of data from some tropical countries concerning economic level, population and education, compared with UK and USA[3]

	Population (millions)	Gross National Product (per capita, $US)	Adult literacy (%)	Number enrolled in secondary school (% of age group)	Number enrolled in higher education (% of population aged 20–24)
Low income countries					
Bangladesh	88.9	90	26	22	3
India	659.2	190	36	28	8
Sri Lanka	14.5	230	85	52	1
Tanzania	18.0	260	66	4	less than 1
Middle income countries					
Egypt	38.9	480	44	47	14
Mexico	65.5	1,640	82	39	11
Hong Kong	5.0	3,760	90	57	10
Industrial market economies					
UK	55.9	6,320	99	83	19
USA	223.6	10,630	99	97	56
Capital surplus oil economies					
Saudi Arabia	8.6	7,280	not known	26	7
Kuwait	1.3	17,100	60	74	13

even in the oil-rich countries to produce sufficient numbers of well-trained and capable technicians to support the local scientists and teachers. There is, therefore, an urgent need for designs of well-considered simplicity, with reliance on high technology kept to the minimum required in specialist laboratories.

EDUCATIONAL LABORATORIES IN THE THIRD WORLD

The most pressing need for laboratories in the Third World is at the primary, secondary and higher levels of education, with a strong emphasis at primary and secondary levels (Fig. 7.3). Much excellent work has been done to encourage the design of science facilities for schools in a way that assists both teachers and pupils to participate in scientific activity in a positive way without making too great demands on national resources. Unesco in particular has supported regional centres which have produced excellent publications for the guidance of teachers and architects. These have concentrated on the design of sturdy laboratory benches and tables using basic equipment made, as far as possible, from materials

Fig. 7.3 — Chemistry laboratory at the University of Juba. Example of a simple, low-cost laboratory in a remote area of the Third World and built in a converted secondary school building.

available locally. It is recommended that finishes should be durable and easily cleaned. Drains for waste liquids in particular should be accessible for cleaning. Laboratory layouts should as far as possible avoid fixed island benches to give better flexibility, with any fixed benches, services and equipment confined to the outside walls (Fig. 7.4). In hot/dry areas ample closed, storage cupboard space is necessary to protect equipment and chemicals from dust, whereas in hot/humid areas all storage space should allow for good through-ventilation.

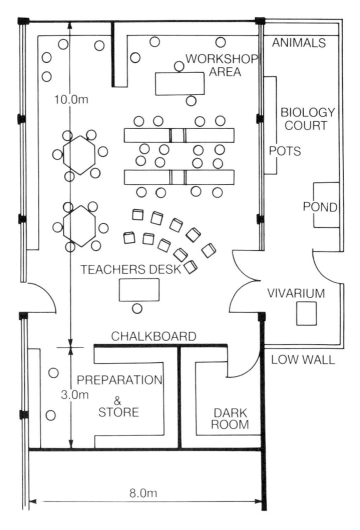

Fig. 7.4 – Layout of a multi-purpose laboratory suggested for use in secondary schools in Asia[4].

Fig. 7.5 – Low-speed wind tunnel in use outside the School of Architecture,
University of Khartoum.

In many countries, simple scientific experiments and demonstrations
may be conducted in the open with the minimum of shelter, particularly if
portable equipment is used (Fig. 7.5). Indeed, if it is accepted that the meaning
of laboratory is 'a place used for scientific experiments', then many community
primary schools in the Third World with school gardens where new plants and
crops are demonstrated to farmers and children alike can be said to have open
air laboratories. Provided shade from the direct rays of the sun is available and
there are no dust-laden winds, quite a range of scientific activities can be carried
out in the open if the environmental conditions are more comfortable than
within an enclosed building.

THE BUILDING FABRIC

Unfortunately, there are many examples of research institutes and university
laboratories where initial over-investment in buildings and equipment has led to
lack of resources for keeping laboratories working to full capacity. Funds for
materials and spare parts are often insufficient and foreign currency for imports

is frequently not available, especially in recent years when high energy costs and low commodity prices have combined to cripple the economies of countries without their own oil supplies.

Rising energy prices have accentuated the need for reduction in energy consumption. Wherever possible passive means should be used to obtain the desired internal environment. In those cases where it is still necessary to use mechanical means of cooling, the energy load should be kept to the minimum by intelligent design of the building fabric. The shortage of foreign currency means that the proportion of imported items must be reduced to the minimum.

Whereas in most countries it is difficult to achieve a small percentage of imports for the supply of scientific apparatus and equipment, it is certainly possible to reduce the percentage of imports more than is usually the case for laboratory furniture, services and the building fabric itself. Even the use of locally produced non-traditional materials should be kept to a minimum because both the raw materials and the fuel used in their manufacture must in many cases be imported.

In the case of metals, the supply of steel bar and other simple sections is usually reliable, even though these must be imported, except in a few relatively self-sufficient countries like India. Metal-working skills are omnipresent, although not necessarily to be found within the formal building industry. Indeed, many skills that can be useful in the construction of laboratories are often found to be developed to a higher level in industries not directly related to construction, for example, the mining industry. Roofing sheets made from aluminium and galvanised sheet steel are frequently rolled and cut to length in local plants although again the sheet metal is usually imported from a developed country, often at a relatively high price. In some cases, however, building materials may be imported from other Third World countries without the need for hard currency; Bangladesh, for example, imports corrugated galvanised steel roofing sheets from neighbouring India.

In every case it can be said that the use of reinforced concrete should be avoided where possible because both cement and steel are costly, energy-hungry, and usually require hard currency for import. There are of course oil-producing countries with no hard currency problems, although there are a few like Nigeria, Mexico and Indonesia who have serious economic problems in spite of being oil producers. Provided that there is sufficient land available, single-storey development is preferable to multi-storey as it avoids staircases and suspended floors which usually require the use of reinforced concrete. The most economic structure is, therefore, single storey with load-bearing walls. The roof-supporting structure should be of timber unless a wide span is required because few countries are without a supply of local timber which can be used in pole from when required, eucalyptus and teak are examples. For walls, burnt clay bricks and blocks may be the most appropriate local choice, whereas some countries are fortunate to have supplies of stone which can be worked economically.

Stabilised soil has been used intermittently for housing since the late 1940s, but has not been used for scientific and educational buildings because of doubts concerning its structural strength and durability, particularly its resistance to erosion. Various stabilisers have been used, including cement, bitumen and lime, as have successful manually operated block-making machines. The United Kingdom's Building Research Establishment is currently promoting a manually operated block-machine with a hydraulic ram to increase the compression of the block. This results in a stronger, durable and more precise product made from selected soil, stabilised with the addition of lime. The machine can also produce flooring and roofing tiles with the use of appropriate moulds. There is now sufficient historic experience of blocks made from soil stabilised with cement to justify confidence about durability provided that there is some protection against erosion from driving rain where this is likely to occur. There is a strong case for extending the use of stabilised soil blocks particularly in hot/dry climates.

Fig. 7.6 – Department of Biochemistry, Faculty of Agriculture, University of Khartoum. The double roof and concrete sunbreakers have been used to reduce the effect of solar radiation. The external form of the building implies a warm climate.

The roof surface is, of course, exposed to the greatest amount of solar radiation and therefore should reflect as much as possible and have minimal thermal capacity. In this respect corrugated metal sheeting gives the best results provided sufficient insulation is provided to prevent re-radiation to the interior from the inside surface which can reach very high temperatures. In hot/dry areas it has been found that the most effective roof construction from the thermal point of view is a suspended, solid reinforced concrete slab with an air gap between it and the outside surface of corrugated metal sheeting, the concrete slab acting as a thermal reservoir (Fig. 7.6).

SUMMARY

When designing laboratories for tropical climates:

(a) Adequate climatic data should be made available.
(b) The design of the building fabric should allow for passive interaction with external climate to give the best possible conditions for internal comfort.
(c) Mechanical means of cooling, if required, should be as simple as possible and with minimum energy consumption.
(d) Imported equipment and building materials should be kept to the minimum compatible with the scientific and environmental needs of the laboratory.
(e) Laboratory equipment, benches, services and built-in fitments should be as sturdy and durable as possible.
(f) Laboratory layouts should be flexible to allow for future change of use.
(g) Laboratories and their equipment should not be reliant on an unrealistic level of technician support and maintenance.

REFERENCES

[1] O. H. Koenigsberger, T. G. Ingersoll, A. Mayhew and S. V. Szokolay, *Manual of Tropical Housing and Building, Part I: Climatic Design,* p. 58, Longman, London, 1974.
[2] M. H. Rahman, *Some Approaches to the Problems of Squatters in Dacca,* M. Phil. Dissertation, University of Newcastle, 1975.
[3] *World Development Report 1981,* The World Bank, Washington, USA, 1981.
[4] *School Building Design, Asia,* Asian Regional Institute for School Building Research, Unesco, Colombo, 1972.

8

Assessing the cost for conversions

W. P. Horsnell

Laboratories Investigation Unit, Department of Education and Science

The initial construction cost of a building is not a true measure of its value in investment terms. To arrive at this the running and maintenance costs throughout its assumed life, must be included. These costs have been disregarded to a large extent until recently. This omission is not necessarily a criticism of designers but stems from the policies of governments over many years where in public sector building programmes, both in Europe and the United States of America, budgetary control is concentrated on initial capital cost which is usually unrelated to acquisition costs or the cost of furniture and fittings, and pays little or no regard to continuing costs.

As a result it is somewhat surprising, and to their credit, that in spite of this designers produce buildings which recognise and have favourable influence over running and maintenance costs. It is the author's view that cost control must continue to be directed at the initial cost but not without regard to the effect of design decisions upon costs arising during the life of the building. The complexities of these life or running costs can be more readily appreciated if looked at as mortgage payments. In other words they are brought to a common base, by discounting, so that their true relationship is established. Figure 8.1 shows a breakdown of these capital and recurrent costs for a science facility of an English university.

Consideration of the relationship between land and building, and adaptations and costs compared with equipment and its replacement, shows that the latter accounts for just over half the total cost. This shows that the user recognises that equipment will become obsolete or wear out but clearly is content to install it and work in an environment which is possibly quite unsuited to the change brought about by the new equipment judged on the expenditure incurred. At first consideration the real effect may not be apparent, but the position changes when staffing is included (Fig. 8.2). Capital costs now shrink to 14 per cent of the total and the relationship between building and equipment costs remains

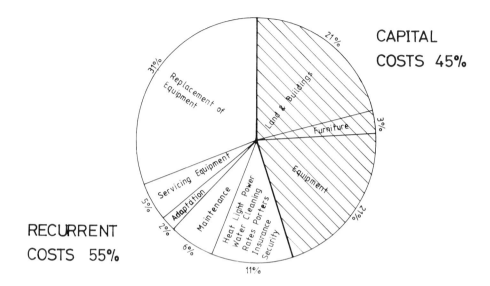

Fig. 8.1 – Capital and recurrent costs for a science faculty of an English university.

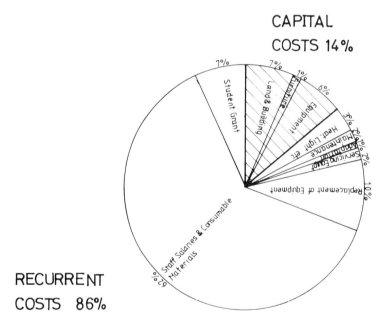

Fig. 8.2 – Capital and recurrent (including staffing and consumables) costs for a
science faculty of an English university.

roughly the same. The question now is whether anything can be done to contain recurrent costs which now stand at 86 per cent. It may be said with some justification that staff costs are beyond the control of the architects who can have no influence on them. On reflection it will be seen that this is not strictly true and that any action that can be brought to bear on the design to minimise inefficient working by the users or their equipment must be implemented. The most important and significant example of this is the ability of a building to respond easily and cheaply to the changes that will be demanded of it during its life. This requirement to be able to adapt to changing needs is probably more pronounced in laboratory buildings than in any other type of building. Such buildings have proved in the past to be very difficult to modify and consequently expensive to change.

The cost of adapting the science facility considered earlier was about nine per cent of the initial cost of the building and its furniture. This represents only the physical work, i.e. the actual building work; it cannot show the inconvenience which may have been endured by the users or changes, or possible improvements foregone to avoid disruption and consequent expense. The cost of disruption which must accompany physical change is extremely difficult to evaluate but that does not mean it has no value. Academic Building Systems, a research group within the University of Indiana, have calculated the cost of disruption to university science programmes in the USA as between 15 and 30 per cent of the cost of the alteration work which caused the disruption. This suggests that the discounted value of the real cost at constant prices of adaptation throughout the life of the building to meet the changing requirements of science almost certainly ranges between 10 and 20 per cent of the initial cost of these facilities. Any reduction in these costs to the point of elimination would be, therefore, a worthwhile exercise.

The normal method of overcoming problems of flexibility in the past was to over provide services. In conditions existing today this is a questionable and almost unacceptable solution as any future savings can easily be swallowed up by extra initial costs. Where does the answer lie? The author suggests that designers must move towards a more flexible approach in laboratory design in which work places and service outlets are not rigidly planned but give the operator mobility within the laboratory and the opportunity to redeploy with the absolute minimum of disruption, inconvenience and cost.

Any approach to design that entails uncertain financial risks must be rejected in favour of one that operates within any normally established and accepted expenditure limit. A decision to invest in prolonging the life of a building must, as with any investment, depend upon the value so secured. In this case this means comparison with an alternative investment, that is a new building. This rests upon the principle that a converted and restored building must not only provide acceptable standards of quality and function but also cost no more than the equivalent new building.

Different approaches to cost appraisal are set out in Appendix 8.1 and they provide a guide to budgets which might become available for conversion work. Appendix 8.2 shows a method of assessing the cost of a conversion project and compares it with the options considered in Appendix 8.1.

SUMMARY

Capital cost whilst very important is not the only criterion to be considered. If that were so the only advantage of conversion would be cost plus quicker possession and occupation. An owner or occupier is entitled to expect more than that for his expenditure. Investment should, for example, be capable of abatement to the extent that parts of the building can be retained at little or no extra cost. The comparison should also embrace site values and limited life expectancy, plus differential running costs and the amenity benefits to society in retaining further use of sound buildings of the past.

APPENDIX 8.1

ECONOMIC CONSIDERATIONS – INVESTMENT APPRAISAL

Example 1 – Purchased building
ALTERNATIVE INVESTMENT

		£
New building of 3,500 m² @ £520/m² gross including external works and furniture & fittings		= 1,820,000
Consultants' fees (12½%)		= 227,500
Site: 0.14ha @ £0.5m/ha		= 70,000
Total investment		2,117,500

PROPOSED INVESTMENT

		£	
Purchase of 3,800 m² building	=	720,000	
Excess running costs 300 m² @ £9/m² p.a. = £2,700 p.a. Present Value (PV) of £2,700 p.a. for 50 years @ 5% compound interest	=	49,300 =	769,300
			1,348,000
Consultants' fees for alterations (20%)	=		224,700
Limit of expenditure on alterations			1,123,500

Example 2 – Rented building

		£
ALTERNATIVE INVESTMENT as before		2,117,500

PROPOSED INVESTMENT

         £

Renting of 3,800 m² building
@ £12/m² p.a. = £45,600 p.a.
PV of £45,600 p.a. for 50 years
@ 5% compound interest   =   832,500
Excess running costs as before   =   49,300 =   881,800

                                    1,235,700
Consultants' fees for alterations (20%) =   206,000
    Limit of expenditure on alterations   1,029,700

Example 3 – Owned building

		£
ALTERNATIVE INVESTMENT as before		2,117,500

PROPOSED INVESTMENT

         £

Market value of owned building   =   587,000
Extra accommodation by extension
of existing building = 500 m² @ £520/m²   =   260,000
Consultants' fees (12½%)   =   32,500 =   879,500
                                    1,238,000
Consultants' fees for alterations (20%) =   206,000
    Limit of expenditure on alterations   1,032,000

Example 4 – Short-life renewal

		£
ALTERNATIVE INVESTMENT as before		2,117,500

PROPOSED INVESTMENT

         £

Renting of 3,800 m² building
@ £10/m² p.a. = £38,000 p.a.
PV of £38,000 p.a. for 18 years
@ 5% compound interest   =   444,200
Excess running costs
300 m² @ £9/m² p.a. = £2,700 p.a.
PV of £2,700 p.a. for 18 years
@ 5% compound interest   =   31,500
Sinking fund
Investment @ 5% compound interest
to produce £2,117,500 in
18 years   =   879,800 =1,355,500

                                     762,000
Consultants' fees for alterations (20%)   127,000
    Limit of expenditure on alterations   635,000

APPENDIX 8.2

METHOD OF ASSESSMENT OF LIKELY EXPENDITURE ON ALTERATIONS

	Normal cost distribution of comparable new building (£/m² gross floor area)	Estimated alternative (%)	Equivalent cost (£/m²)
Substructure	29	—	—
Superstructure			
External walls	37	—	—
Upper floors	19	—	—
Roof	28	—	—
Stairs	3	—	—
Windows	10	40	4
Partitions and doors	21	50	11
Finishes			
Walls	10	20	2
Floors	15	50	7
Ceilings	12	50	6
Furniture & Fittings	98	85	83
Services			
Sanitary fittings	34	70	24
Cold water	30	33	10
Hot water	11	33	4
Heating	23	15	4
Fume extract	22	70	15
Gas and other services	16	33	5
Electrical	55	40	22
Drainage	9	50	5
External works	38	—	—
	520		202

Extra cost of working
in existing building,
stripping out and
unforeseeable work

ADD 15% =   30

Total likely cost   £232

£

Total likely cost = 3,800 m² @ £232/m² =   881,600

Add Consultants' fees (20%)   =   176,300

Total likely expenditure   1,057,900

Since the figure of £1,057,900 includes consultants' fees it must be compared with the corresponding figure given by the investment appraisal (Appendix 8.1). Thus, if it were assumed to apply to any of the foregoing costed examples, it would compare as follows:

Example 1: It is 22% less than £1,348,200 and therefore favourable.
Example 2: It is 14% less than £1,235,700 and therefore favourable.
Example 3: It is 15% less than £1,238,000 and therefore favourable.
Example 4: It is 39% more than £762,000 and therefore unfavourable.

9

Building to a cost – a planned approach

B. G. Whitehouse
Laboratories Investigation Unit, Department of Education and Science

INTRODUCTION

Cost in the context of this chapter means the initial capital cost of providing the building and of its associated site development – fencing, planting, surface treatment, roads, paths, service mains, etc. Other capital costs will be involved and these will include:

(1) Professional fees and charges – the services of professional designers and consultants, site supervision by one or more clerks of works. These will flow from the building costs.

(2) Loose furniture and equipment – the items required to bring the development into use after completion. This is additional to and overlaps provision in a normal building contract.

(3) Site – either purchase, leasehold interest in it, or rental of existing buildings.

Site costs are important when determining the appropriate cost of converting an existing building and this raises the general issue of present and future costs and how to put them on an equivalent basis. A freehold interest is usually unencumbered and is a simple present cost. Leasehold interests, however, are for a short or long life, and will bear annual charges such as ground rent and possibly a full repairing covenant or some lesser covenants. Rental interest would be a regular periodic charge including some recurrent costs such as rates, but excluding others such as energy use. The various costs – present or future lump sums or regular payments at various periods – must be put on an equivalent basis for comparison of alternatives.

It is desirable, when considering the initial building provision, to examine proposals of spending initially in order to save later. Examples include:

(a) Aluminium window frames which require no future decoration costs compared with timber frames requiring external scaffolding and painting every five years.

(b) Double glazing to conserve heating energy.
(c) The provision of working surfaces, storage requirements and servicing of
     laboratories that provide future flexibility and adapatability and consequent
     saving in refurbishment costs.

These are a few examples of wider concept in setting a cost target for a project –
the consideration, as far as this is possible, of total life costs or relevant parts of
them.

The consequences of accepting such features does not necessarily entail an
increase in the normal unit cost for such accommodation, since standards of
reasonable energy conservation and quality of the building fabric are already
incorporated. The issue is more one of obtaining value for money in space
demands and building geometry, and of a rigorous examination of component
costs to ensure adequacy but not over-provision.

The costs incurred at different times during the design and construction of
laboratories are equated by discounted cash flow techniques, examples of which
and a bibliography for further reading are given in a Treasury booklet[1]. For
low-risk investment, such as energy-saving schemes, use of a 5 per cent discount
rate is suggested, and for high risk proposals, 7 per cent. This is a long-term
economic view of investment or commitment to future costs and is probably in
conflict with an accountant's or financier's view of the future value of money.
The latter would be far more cautious, choosing a rate nearer to the prevailing
cost of borrowing. The lower the interest rate the greater the effect of future
costs or savings, yet these are less certain than present costs. In the author's view
the key is to take a realistic long-term rate of interest on borrowing, be cautious
about future benefits and slightly pessimistic about future costs.

The broad economic assessment of alternatives to the provision of laboratories, consideration of future capital and costs and the approach taken towards
the design are all very relevant issues. These will all determine the appropriate
cost target and this chapter assumes that this has been decided already.

THE APPROACH TO DESIGN

Many items in a laboratory will have a limited life either in terms of their
physical function or of their obsolescence in use. Motors of all sorts will break
down and will have to be repaired, refurbished or replaced. New equipment, not
in use at the time a laboratory was designed, will require a fresh consideration of
the use of space, as well as the furniture, storage and servicing requirements of
the equipment and facilities supporting its use.

A period of not greater than 20 years will probably be the most suitable
base for economic appraisal of a proposal. Within this time motorised equipment
will have been replaced or had a major overhaul, loose furniture will be due for
replacement, the floor finishes and electrical installations might be on the verge

of renewal. the tasks performed in the laboratory will almost certainly have changed substantially, perhaps necessitating more than one process of adaptation. Technological changes in building and servicing standards will probably require upgrading work.

The Laboratories Investigation Unit has since 1971 published a number of reports[2] which are appropriate to particular circumstances. These projects have shown that all laboratories grow and change although some are more subject to this than others. Designing for this involves not only relating structure, space, services and fittings but also deciding which elements shall be fixed and which could be changed. Other design considerations involve minimising the demand for space which costs money to heat, light, clean and maintain and will incur payment of rates.

Once a site or building for the laboratory has been selected and alternative designs explored, together with consequent recurrent spending, then the design team has a brief incorporating the selected design features. To arrive at the target cost, such a target must be set in the brief and include a target of area and of unit cost. Table 9.1 illustrates the range of areas appropriate to educational facilities and to these must be added staff social and dining provision, administration, library facilities, etc., for which there are other area norms. Associated with the gross floor area for the building are unit costs which can very from £300/m² to over £500/m² for more intensively and extensively serviced laboratories. The increase in cost over this range will usually be almost wholly due to furniture and services although particular requirements for building form such as high ceilings will sometimes play a part.

The target cost will be divided amongst the elements of the building, agreed by the design team as a whole. This cost must include allowances for inflation to completion date, a contingency sum to meet unforeseen problems during the construction phase, abnormal site work problems, such as ground bearing pressure, a high water table, excessive slope of the site, and a notional allowance for site works. The sum of these costs will form part of the brief to the design team. A simplified example of a cost plan is shown in Table 9.2. If required by the design concept, the normal allowance for loose furniture should be incorporated in the cost target at this stage and the normal elemental costs for services adjusted accordingly. The element for built-in furniture and fittings in Table 9.2 reflects the traditional provision for fixed working surfaces and storage in laboratories.

THE COST PLAN

The cost plan at this point has been drawn up before the commencement of any design work, but will be based on considerable experience. This will include statistical data, in the form of cost analyses for similar projects, for elements peculiar to the type of building being planned and from general cost analyses for

Table 9.1

Space standards for educational laboratories

Educational sector	Source document	Space area (m²/workplace)				Relevant group size
		Laboratory	Ancillary + storage	Balance[a]	Total	
Schools	Curriculum schedules[3]	2.70	0.41	1.24	4.35	30
16–19 years (non-advanced)	Design Note[4]	3.00	0.45	1.38	4.83	15
Polytechnics & advanced	Area recommendations[5]	3.25	3.50	2.70	9.45	25
further education		4.00	3.50	3.00	10.50	15
Universities						
Undergraduate:						
Biology (general-purpose)		4.00	2.40	3.20	9.60	
(biochemistry & experimental)		5.00	3.00	4.00	12.00	
Physics		5.00	2.25	3.63	10.88	per student/worker
Chemistry		5.00	2.50	3.75	11.25	
Postgraduate research		9.50	4.28	6.89	20.67	

a Balance represents the addition of area for general circulation space, sanitary and cloaks provision and general services and cleaning storage spaces.

Table 9.2

Simplified cost plan for a chemistry building

Element	Accommodation cost (£/m²)		Total (all buildings)
	Heavily serviced	Lightly serviced	
Substructure			33.00
Roof			
External walls			} 85.00
Windows and external doors			15.40
	207.30	207.30	
Internal walls and partitions			
Internal doors			} 37.40
Wall finishes			
Floor finishes			} 36.50
Ceiling finishes			
Fittings and built-in furniture	19.00	6.00	11.98
Sanitary appliances, sinks, etc.	7.68	4.22	5.81
Disposal, wastes etc.	39.21	1.80	19.01
Water	36.96	8.70	21.70
De-ionised and distilled water	4.02	–	1.85
Heat source	17.40	17.40	17.40
Space heating	18.38	18.38	18.38
Ventilation	–	4.84	2.61
Fume cupboard extract	25.26	–	11.62
Fume cupboard make-up	25.80	–	11.87
Electrical services	85.94	14.81	47.53
Gas services	14.84	0.54	7.12
Protective installations	1.54	0.65	1.06
Communications	–	1.61	0.87
Compressed air	4.02	–	1.85
Goods lifts	11.54	–	5.30
Builder's work to services	29.65	6.04	16.90
	548.54	292.29	410.16
Drainage	11.00	5.00	7.76
	559.54	297.29	417.92

Contingencies	2.5%		10.45
External works	7.5%		31.34
Abnormals	5%	} Allowances to be justified	20.89
Loose furniture	20%		83.57
			564.17

elements which are necessary whatever type of building is to be provided. The members of the design team will also have a good general knowledge of particular aspects of the building type.

Associated with each element will be elemental unit quantity which is the parameter that reflects the cost pattern of that element, and a connected elemental unit rate. Examples of unit quantities would be:

Frame: steel (tonnes) — derived from previous buildings.

Windows, external walls, partitions: superficial areas — related to energy and ventilation standards.

Doors, lavatories: number.

Electricity, lighting: illuminance level(lux).
            power: number of points and terminals.

Ventilation: volume of air handled/m^3.

Heating: installed load (kW) — derived from energy target.

The quantification of these unit quantities will come from the brief or a development of it. They are standards which are design targets and it is essential that they are determined before commencing the design stage.

Preliminaries, which are included under 'Accommodation cost' in Table 9.2 would, together with contingencies, be shown separately. Also, at this cost plan stage a design and price risk of about 5 per cent would be included, to be used for balancing out the elements as cost checking proceeds.

The architect is now able to prepare sketch plans for approval by the client and, if necessary, adjustments can be made to the cost plan to contain his proposals within the target. Detailed design of the various elements can proceed as soon as the client has approved the proposals. The quantity surveyor will need to check the cost of each using current cost information and preliminary quotations for the more specialised work. Some costs will exceed those in the cost plan, whereas others will be lower. The cost check should, however, again contain the cost within the target, with the designers jointly taking responsibility for any necessary decisions required to achieve this. The cost plan figures for abnormals and site works would also be checked at this stage and the cost plan amended if necessary.

Documents for tenders should incorporate all working drawings, specifications and prime cost items for specialist supply or installation. Any necessity to resort to provisional quantities or documents should be strenuously avoided. It is at this stage that many post-tender problems can be overcome. The tender itself should be a virtual formality. Provided that cost checking has been carried out rigorously, all corrective decisions will have been made and incorporated into the documents. The only significant factor that could upset this state of affairs is the market conditions affecting the construction industry. There was a tendency within the public sector at one time, for tenders to peak towards the end of March and this led to high tenders. It is hoped that this situation has been overcome by the introduction of controls on expenditure rather than on the start of construction. It is as well, however, for the team to have possibilities for savings should a tender be higher than the target cost. These proposals can be put to the client for consideration if the necessity arises.

A firm procedure should be established for post-contract cost control. The architect only should issue instructions which should be costed, and if possible agreed with the contractor or subcontractors, by the quantity surveyor. The client should receive a cost statement together with each certificate he is asked to honour, setting out the contract sum, agreed variations, contract cost fluctuation (if applicable) and the projected final cost. Any claims on the contract, due to exceptionally inclement weather, etc., should be determined as soon as is practicable and included in these statements.

REFERENCES

[1] *Investment Appraisal in the Public Sector: A Technical Guide for Government Departments,* HM Treasury, London, 1982.
[2] *Laboratories Investigation Unit Bulletins,* Department of Education and Science, HMSO, London, 1977 onwards.
[3] *Area Guidelines for Secondary Schools,* Design Note 34, Department of Education and Science.
[4] *Area Guidelines for Sixth Forms, Tertiary and NAFE Colleges,* Design Note 33, Department of Education and Science.
[5] Design Note, Department of Education and Science, in press.

10

The efficient design and construction of a new laboratory

W. R. Tully and **R. F. Young**
Roussel Laboratories Ltd, Swindon

HISTORICAL

The construction of a new manufacturing complex for Roussel Laboratories Ltd was started in 1968 on a site on the outskirts of Swindon. In 1969 it was decided that a United Kingdom research group would be formed and housed on two floors of the building originally designated for offices. Although the accommodation was adequate, the laboratories were not ideal particularly those to be used for chemistry, where the net working area was 11.4 m^2 per scientific worker, with no provision for additional equipment. The British Standards Institution recommends[1] that depending on the nature of the work, 100 ft^2 (9.3 m^2) is likely to be cramped whereas over 250 ft^2 (23.2m^2) is seldom necessary. Ferguson[2] suggests that 180 ft^2 (16.7 m^2) is sufficient. As awareness of potential laboratory hazards and safety requirements increased it became evident that a remedy was required. Therefore, in 1980, after considering all possible alternatives, plans were proposed for a new building to house the Chemistry Department.

THE SITE

The Company's manufacturing site is subject to a long-term master plan in which there is no provision for an additional building for research. The first task was to identify an area which could be allocated to this project without prejudicing the development of production and warehouse facilities.

The best area available was a space occupied by a redundant sprinkler water storage tank. This presented some planning questions due to a major underground sewer, existing roadways, proximity to the site boundary, and a site which was rather narrower than that which would produce an ideal solution.

It was proposed that the building should be designed to accommodate 18 bench chemists with ancillary staff and that high priority should be given to working environment, that is laboratory lay-out, design of ventilation, siting of fume cupboards and safety features consistent with current legislation and recommendations. During the outline planning phase the need for a new clinical pharmacology department was identified by the Company and because of the limitations on site development it was decided that this unit should be incorporated into the same building. The final plan was for a three-storey building with clinical pharmacology on the ground floor and chemistry on the upper two floors.

The Borough of Thamesdown encourages the development of high technology industries in the Swindon area, but when the original proposals were discussed with the local authorities questions from the local community were anticipated due to the proximity of the proposed building to the site boundary. After much discussion it was decided that providing that a reflective cladding material was used and the established standard of landscaping was maintained then planning approval would be given.

THE PROJECT TEAM

Costing a building project may be approached in two ways, with either a fixed limit or an estimated cost based on an agreed specification.

It was necessary first to establish the approximate cost of the desired facilities before proceeding with the detailed planning and costing of the project. To this end a team was set up to examine the requirements of the Chemistry Department and a list of objectives drawn up. At that time it had not been decided which route to follow with regard to the design of the building. The choice was between either traditional use of architect, quantity surveyor, mechanical and electrical consultant, or the use of design and build contractors. The Company normally employs the former method, but the relative merits of the two approaches formed the subject of much debate.

Discussions were held with several design and build contractors and with a quantity surveyor concerning the outline project. They were asked to give an approximate cost of such a project and this revealed that it would be worth proceeding to the next stage of design.

Although proposals from two of the design and build companies were impressive, it was decided to use the traditional method of employing professional consultants to assist with the Company's plans. A team consisting of an architect, a structural engineer and a quantity surveyor was employed, and instead of a mechanical and electrical services consultant, a group of local contractors previously used by the Company was engaged. The authors liaised very closely on detailed scientific requirements and the professional team carried out the design and construction elements of the project.

Table 10.1 shows the cost plan produced by the quantity surveyor. The estimated cost of a building of 692 m^2 was £827.5/m^2.

Table 10.1

Cost plan for new research facility for Roussel Laboratories Ltd

	£
Foundations	17,441.54
Structure	50,567.95
External walls	87,816.88
Roof work	7,901.90
Internal walls and doors	16,989.40
Internal finishings	36,094.01
Fittings	57,550.00
Services	206,924.90
External works	2,001.00
	483,287.58
Preliminaries	42,759.00
	526,046.58
Overheads and profit 5.5%	28,932.56
	554,979.14
Agreed lump sum addition for main contractor's fixed price: provided that date for possession is on or before 24 August 1982	11,500.00
	566,479.14
Contingency sum	6,150.86
Contract Sum	572,630.00

The contract was let on a competitive tender based on a complete design. This method was selected to accelerate the project design period and to ensure that the project team was able to work closely with the selected contractor in determining the most economic method of forming the structural elements and finishes for the building. Wimpey Construction plc were the successful tenderers and were subsequently engaged as the main contractor under a Joint Contract Tribunal form of contract on a fixed price basis over a 38-week programme.

STRUCTURE OF THE BUILDING

The building was constructed in reinforced concrete and because of the nature of the ground it required a piled foundation. A proprietary roofing system was used for the fairly steeply pitched roof. External walls are of toughened glass cladding with double glazed windows in aluminium frames. Internal walls are of conventional plastered blocks finished with paint. Ceilings are of the suspended type using perforated metal tiles in the main laboratories and mineral fibre acoustic tiles elsewhere. Internal doors are of good quality plywood with a paint finish and are fire-resisting. The main access to the building is through two sets of double glass doors the inner of which is fitted with a security card-key locking system. This leads into a central stair well containing a small goods lift. At each end of the building is a metal fire-escape, access to which is gained on each floor by push-bar fire doors.

INTERNAL LAYOUT

Having established that the available floor area for chemistry was 462 m^2, the next task was to determine the optimum amount of space per scientist and how much space could be allocated to specialist facilities and storage. For the type of pharmaceutical research to be undertaken it was decided that each chemist should have a 1.5 m fume cupboard, a 3 m bench run, a 1 m desk and a working area of 17 m^2. This decision was supported by reference to other new organic chemistry laboratories.

The final design incorporated three large laboratories, two on the first floor and one on the second floor. The layout (Fig. 10.1) for each of these laboratories is identical, allowing freedom of movement with no dead-ends and a choice of three exits, one being through the project leader's office into the adjacent office. The fume cupboards are arranged along one wall so that movement parallel to the cupboards and therefore disturbance of air flow is minimised. Location of facilities such as flame-proof solvent cabinets, refrigerator and a comprehensive safety shower unit incorporating overhead shower, drench-hose and eye-wash with aerated water was carefully considered for ease of access. The under-bench units and fume cupboards are of all metal construction, the working surfaces of ceramic sheets bonded to plywood and the instrument benches are covered in formica, all supplied by Morgan and Grundy plc.

Services to each bench unit and set of fume cupboards have independent isolation valves for vacuum, air, nitrogen and water, and an earth current leakage trip device for the electricity supply. The latter feature is extremely useful for identifying faulty electrical apparatus and is also activated, for example, by spillage of water on to the heating element of a mantle. In addition, each laboratory has two master electricity cut-off buttons, an alarm button to call for local assistance and a fire-alarm, linked to the Company's overall site system, which is activated by smoke and heat detectors.

The remainder of the second floor is occupied by an office, which also serves as a small meeting-room, and ancillary rooms. The chemical store contains mobile racking to increase utilisation of limited space and three large refrigerated cabinets cooled by units located outside the room; the lighting is flame-proof and there is no suspended ceiling, to maximise air volume. A separate laboratory houses specialist apparatus, the gases for which are piped in from an external cylinder store, and possesses the same safety features as the main laboratories.

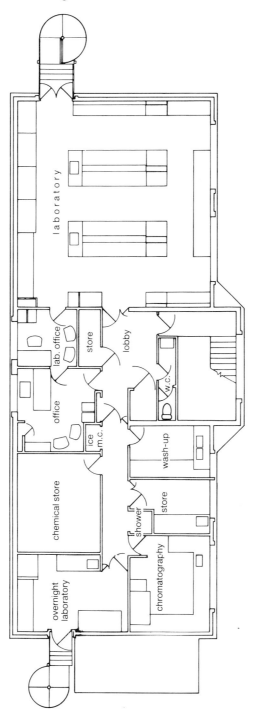

Fig. 10.1 – Laboratory plan. Roussel Laboratories Ltd, Swindon.

Finally, there is an nmr room, a washing-up room and a chromatography room in which air is extracted at floor level to prevent build-up of the heavier solvent vapours. The flooring in all working areas is covered with non-slip, welded vinyl sheet. The landing areas on both floors provide limited toilet and hand-basin facilities, drinking fountain and coat cupboard. There is no necessity for storage facilities for equipment and apparatus additional to that provided in the laboratories.

With the agreement of the insurance company a sprinkler system was not installed because it was thought that there are certain operational dangers associated with accidental activation of such a system in a chemical laboratory. There is, however, a comprehensive supply of fire-extinguishers of all types.

VENTILATION

Figure 10.2 illustrates in simplified form the air movement system for the building. A 100 per cent fresh air supply is introduced at ground level, passed through a steam heat exchanger and humidifier and enters the large laboratories through perforated metal ceiling tiles via a plenum chamber to ensure gentle diffusion. Air flows along the length of the work benches and is extracted through the fume cupboards lined along the wall on the opposite side of the room. The exhaust rate is designed to maintain a face velocity of 0.5 m/s at a sash height of 0.5 m, in keeping with current recommendations.

Due to the location of the building with respect to neighbouring buildings, the height of the exhaust stack was carefully considered in view of approval by the Local Authority and adherence to the accepted limits and recommendations [3]. The exhaust air is taken to a high pressure fan at the base of the stack and discharged at high velocity (approximately 10 m/s) at a level 2 m above the roof line.

Fume cupboard extract is reduced to half speed outside normal working hours. This is indicated by a red light, in the ceiling of the main laboratories, connected to the switch-over circuit. Exhaust malfunction also activates the warning lights through a rotary speed sensor. At the moment, only four fume cupboards have been installed in each laboratory to satisfy the requirements of current staff. Since all the necessary services are in place it will be an easy matter to add additional cupboards if needed at some time in the future. At the same time a heat exchanger could be added to the system to conserve energy; provision for this has been made within the ventilation trunking.

CONCLUSION

The building was scheduled for completion in June 1982 but for various reasons, including a severe winter, it was not ready until October. Financial control throughout the construction period was extremely tight and the final cost was

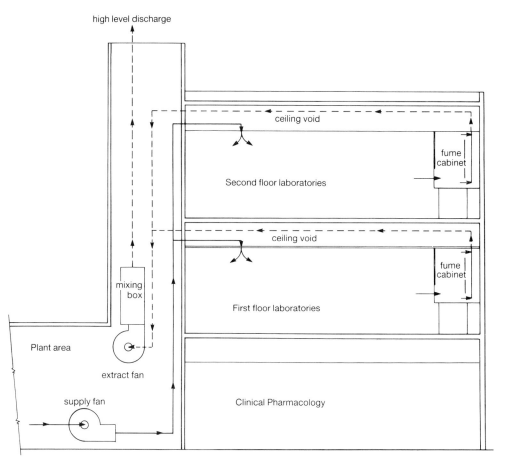

Fig. 10.2 – Diagrammatic layout of air movement system.

very close to that predicted. Now that the facilities have been in operation for more than a year it can be said that the project was an unqualified success with regard to providing a safe, pleasant environment for chemical research. This is due to the close in-house monitoring of the project from conception to completion.

Clearly not all organisations are able to support a full-time project team, but whenever possible the design of specialist facilities should come from an in-house source as both briefing and monitoring the work of consultants takes up a disproportionate amount of time. The Roussel Laboratories' team believes that it has been able to provide a first-class laboratory by paying attention to detail and ensuring that the bulk of the finance is allocated to scientific requirements while keeping the frills to the minimum.

REFERENCES

[1] BS 3202:1959, *Recommendations on Laboratory Furniture and Fittings,* British Standards Institution, London, 1959.

[2] W. R. Ferguson, *Practical Laboratory Planning,* Applied Science Publishers, London, 1973.

[3] D. Hughes, *A Literature Survey and Design Study of Fume Cupboards and Fume Dispersal Systems,* Science Reviews Ltd, London, 1980.

11

Project management in laboratory design

J. H. Armstrong
Building Design Partnership, London

INTRODUCTION

Management is about people. One of the great creative delights of managing a project is being able to co-ordinate and release the talents and enthusiasms of a variety of different disciplines. These talents and enthusiasms, particularly the latter, seem very often to be points of conflict between members of the team and the way in which they work. Some general ideas about the relationships between those involved with research and development and those involved with the execution of projects are considered in this chapter.

PLANNING A NEW FACILITY

When setting out to provide a new facility for science and research it is first of all necessary to decide what is needed, and then what resources are needed to satisfy that functional requirement. The resources are various. People are obviously a prime resource — challenging, dedicated, people willing to work. They must have skills, as people by themselves, even with the best of enthusiasm, are not much use without the requisite skills. They need space in which to work; they need equipment with which to work; they need finance and, crucially, they need time. Time varies in quanta, and seems at times to be rather elastic, but it is an essential component of any management exercise. These, therefore, are the resources needed: people, skills, space, equipment, finance and time.

It is difficult to get access to these resources. Most of this takes place before teams are engaged and facilities designed. It is at this time that laboratory directors are very busy trying to persuade their fiscal masters that the facility is essential and that the resources should be provided.

Once access to resources is obtained, the next step is to consider how to procure the facility; and this, first of all, requires a detailed brief. Having got a brief, design of the facility can begin. There is a very large number of different activities encompassed within that very broad description 'design'. The organisation, assembly, and management of the design team will be mentioned later.

The next stage is to construct the facility. Again the description 'construction' is rather a broad one. It may mean major modifications to an existing establishment or it may mean a completely new facility on a green field site. All the equipment must be procured and, apart from the finance, obtaining competitive prices and tenders, and deliveries of the highly specialised equipment when it is needed is a very skilful and time-consuming business.

Having procured the equipment and constructed the facility, everything must be tested. Finally, the whole facility must be commissioned and the staff trained.

ASSEMBLING THE PACKAGE

Monitoring the procurement process is really where project management comes in. It means putting together the whole package, making sure that things are done when they are supposed to be done, that they are contained within the budgets that are set, and, above all, that they provide the facility which is needed. Monitoring the procurement is therefore concerned with the four factors of quality, quantity, time and cost. There is little point in procuring a facility within the budget and in the time required if the quality of the work or the scale of the facility does not meet the original need.

Staff will comprise the administrative staff for the whole facility, the professional scientific staff, the technician staff and the maintenance staff. All these have to be considered and put together in the right balance related one to the other, and much training is required as well. Therefore the management of the operation to provide the necessary staff resources must proceed in parallel with all the other activities that are taking place in connection with the project.

There is an important aspect of managing quality projects, which is different from managing normal building constructions, and this is the client body. Laboratory clients, rather like hospital or academic clients, are difficult people to deal with. The main reason for this is fairly simple. Scientists are concerned with giving a continuous very high quality service to society. Their main aim is the discovery of truth on the health of society. They do not want particularly to have that kind of activity interfered with by people who insist on asking questions and wanting decisions.

THE COLLABORATIVE PROCESS

The day-to-day running of a research or testing laboratory organisation is a continuous process, every day presenting problems which are dealt with as far as possible as they arise. Research continues and scientists, working probably in fairly small teams, pursue particular aims. The development in their science around the world is monitored to ensure that they are up-to-date and then, sooner or later, a point is reached where they decide that major extensions are needed to the facilities available. This produces what is called 'a project', and that arises somewhere below the horizon of their general work. It may have been turning over in the minds of the facility and management for a few years as they become more and more dissatisfied with the facilities they have and want something better. It is then that a commitment is made. They find they have some access to resources and things start. That is project execution. It seems quite reasonable at first and consists of a few board meetings. Then it begins to take over and grow, and can overwhelm the performance and the facilities available in the existing institution. This presents problems, not least of which are the kind of people to be dealt with (Fig. 11.1).

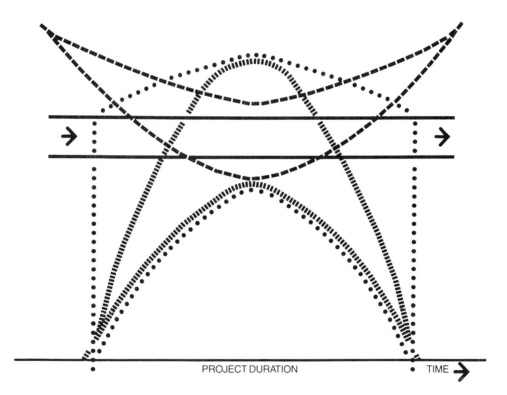

Fig. 11.1 – Project management relationships. Key: ▬▬▬ financial control; ─── continuous facility management; ■■■■■■ finite project execution; ●●●●● project management.

The scientists and their administrative, maintenance and technician back-up suddenly find they are having to interface with the 'builder people' who insist on asking questions which require the scientist to think about past and future practices rather than just getting on with it. These 'builders' insist on having decisions by a certain date and generally begin to interfere. Work has probably started on the new facility and there is some sort of run-in period during which the management considers what is wanted and they may at this point begin to talk to a design team about putting together their facility. It is here that the interface between the facility managers and the project execution team starts and continues.

This collaborative exercise has to take in the whole of the project execution and relate to the facility managers as the job proceeds. There is, therefore, a very important interface between the different kinds of people. The most important task of any project management team is to find people able to overcome these differences.

MANAGEMENT OF THE PROJECT

Experience gained by the Building Design Partnership (BDP), from a large number of major laboratory projects it has undertaken, suggests that the selection of the right people for the project team is of critical importance. The approach taken by BDP in project management is to consider how best to provide a service to the client which interferes as little as possible with the client's work. The quality of the staff required is very important. The first thing to be done is to take away the teams, lock them up together for a couple of days and let them talk about what they do. At the end of that period, people from the facility management and from the design team are selected to form the liaison nucleus for the whole project. What matters is that they understand each other and sympathise with each others' work.

The design team is interested in executing the project and finishing it on time at minimum cost. It will finish the project, settle the final account, clear a few snags, answer a few questions for a year or two after it is finished, and then, if it has done its job properly, leave. Its aims are finite, whereas the facility management team's aims are continuous. This produces quite a difference in mentality. The designers are trying to get something finished, the scientists are trying to make something continue. If that difference is recognised then there will be a different approach to obtaining the brief from the facility managment team which will relate to the people concerned in that team.

Facility managers must recognise that whatever happens, regardless of how good the appointed people are for the liaison work, some time is going to be required from the laboratory managers and the scientists themselves to provide the information needed by the project execution team. This must be allowed for in their programmes, either by increasing staff and maintaining the effective manpower available for the scientific work, or by consciously reducing the amount of work that is to be done. Increasing the size of the facility management team would mean temporary recruitment of staff. This is not very effective because there must be a run-in period of several months, if not a year or two, in most research and development institutions before new staff really can be in a position to take over. The whole relationship between the facility management and the project management team must be very close, and there must be two or three people in the combined teams who thoroughly understand all that is required on both sides.

What is needed is one project liaison officer with sufficient wisdom to identify in the teams those people best able to relate to one another. There may be, for instance, a services engineer in the design team who could relate very closely to the engineering support staff for the laboratory facility. If there are two people who get on well together and understand each other, this should be encouraged and the placing of a barrier of one man between them avoided. Line management can be retained, but communication must be on a broader front. This helps the whole operation to become rather more stable and more conducive to good results.

When projects begin there is an important factor – the paymaster, the financial controlling body, which may not be the actual facility management. In the private sector it might be a main board of directors, it might be a Ministry in the case of a public sector work, but either way it begins to take an extraordinary interest in the project. Armed with a whole lot of awkward financial questions and control systems, it demands value for money at the right time and in the right place. The financial controllers have therefore, also to be taken into the whole project management exercise.

CASE STUDIES IN PROJECT MANAGEMENT

The management structure used for three laboratory complexes designed by BDP provide examples of project management in practice. Two are from the public sector and one the private sector for a large-scale industrial establishment. Whilst each of these projects has apparently different management styles, when the different management structures are examined they all have quite a lot in common. There is a project executive team of the kind previously mentioned; there is a facility management team; there is a project management link of some sort between them; and there is an external paymaster on all of them.

The basic requirement of all these projects is for a project manager, a team leader, a design team leader, quantity surveyor, mechanical/electrical engineers, architects, civil/structural engineers as part of the design team, and a main contractor served by subcontractor and suppliers. This is the project executive team.

In the facility management team, there is a facility director. There is liaison with users, probably through a staff member appointed from within the facility management team. There are unit technicians, there are the heads of perhaps different kinds of research units within the facility and there are specialists who would need to identify their needs to the executive team. There is also an equipment officer responsible for identifying, obtaining tenders and procuring equipment. These last functions are probably performed already within the facility management team or provide a service to it, but resources will need to be scaled up in order to meet the increased needs otherwise the equipment officer and the working systems are likely to be hopelessly swamped by a very large multi-million pound project.

It is interesting to note how this system operates on the different projects. On a Public Health Laboratory Service (PHLS) project at Colindale, there was a project manager who was not part of the design organisation, but separately appointed from BDP. The quantity surveyors, service engineers, architects, civil/structural engineers were independent consultants appointed specifically to the project executive team and in this case the contractor formed a consortium with the major services contractor to provide a joint venture package for the project as a whole. In the facility management team there were unit heads, unit technicians, and the departmental deputy head who acted as the effective co-ordinator of the facility team. There was also a PHLS board to whom reports were made, and a capital project committee on behalf of the Department of Health and Social Security. This case fits into the broad pattern of a project executive team, facility management team and a project management linking the two. On another PHLS project, BDP acted as the design team with an independent project manager. this project had a structure similar to that used at the Colindale laboratory.

On a large organic chemistry research building for a major industrial plant there were internal project managers, not from within the facility team but from within the client body. BDP acted as project executive designers in this case. The management from within the client body was found to work well. The pattern is slightly different, the names are slightly different, the relationships with the projects manager to the client body differ, but basically the functions to be performed are the same.

The management diagrams produced in each case by the project managers looked quite different, but the essence of each was a basic structure (Fig. 11.2). The differences between the kind of people that make up the facility management team, and the kind of people that make up the project executive must be considered.

CONCLUSION

The project management organisation is not just a group of people who have a particular facility for drawing graphs, or monitoring records, or plotting

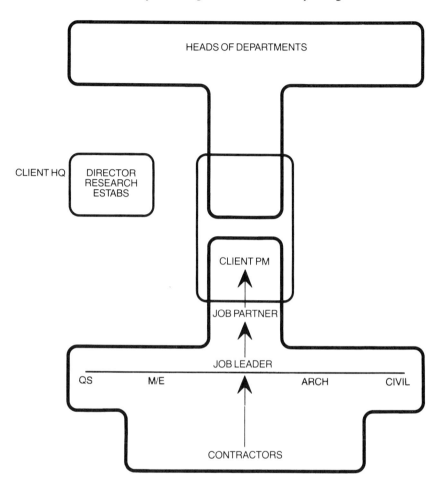

Fig. 11.2 – Operational groups.

expenditure, or keeping a record of variation orders, or preparing minutes of meetings. The person required for project management must be not only a good administrator, but somebody who understands the important and creative temperamental differences between the facility managers and the project managers. If those temperamental differences are harnessed and brought together, then there is an opportunity for a creative development using the thrust towards the completion of a project available within the project executive team, together with a real understanding of what is done in the facility and the desire to provide a continuing service. These together can produce a first-class project which provides the sort of continuing and flexible facility that is needed by the user.

12

The working environment – improvements to meet user needs

M. E. Peel
Glaxo Group Research Ltd, Ware

INTRODUCTION

The total impression that a laboratory worker has of the environment is the summation of various factors which include an appreciation of:

(1) The general environment – the advantages of the town and county in which the laboratory is located; the choice of places to live; the amenities available for the family to pursue their hobbies and interests.
(2) The Company environment – the benefits which, while not directly concerned with work, are available to employees. A Sports and Social Club and a staff restaurant are examples.
(3) The working environment – the environment where the work is performed.

This chapter considers some factors which can be designed into a laboratory, or into a building of which the laboratory forms a part, which have a direct influence upon the quality of the working environment. These factors can be categorised as those which are creature comforts, those conducive to good work and others which are essential for specialised tasks.

A person is comfortable if the temperature and humidity of the surrounding air are within certain limits. The air needs to be fresh, free from smells and dust, and moving gently. The deviations from ideal that can be tolerated vary from person to person and will also depend upon the particular tasks that have to be carried out. The amount and kind of noise at the work place, and the quality of both the general and any specific lighting that is provided affect comfort, and people will have personal preferences.

Within a small to medium-sized industrial building, not much more needs to be done to counter changes in the weather than is now common in many homes. Central heating, double glazing, sun shading, insulation and simple ventilation may be installed. The difficulties of distributing heat and moving fresh air increase with the size of the building and reach a point that demands a good air conditioning system. If the building is to include laboratories, the air moving plant, the ducting, the heating and all the control systems must be carefully designed. If some of the laboratories require fume cupboards, the arrangements for extracting and supplying air have to be integrated with the requirements of the fume cupboards and the whole system balanced using an auxiliary air supply.

The resources needed to operate an air-conditioning system are air, water and energy. In a complex building the conversion of raw air to a condition of chosen temperature and humidity and the distribution and extraction of this conditioned air will also be complex. All components of the system will need to be kept in balance, preferably by a ventilation engineer. Conservation of energy is now very important; optimum operation of air conditioning in a complex building requires that the equipment be computer controlled.

THE GLAXO GROUP RESEARCH LABORATORIES AT WARE

Glaxo Group Research has built three major laboratory buildings on a single site at Ware in Hertfordshire during the past twelve years. The Biological Research Laboratories were completed in 1972 at a cost of £750,000 and are partially air conditioned. The largest building was completed in 1979/80 at a cost of £12 million and houses chemical, analytical and pharmaceutical research, the biochemical, pharmacology, medical and administration departments. This building is fully air conditioned and mechanical and electrical plants are controlled by a Honeywell Delta computer. The third building was built specifically for long-term toxicity studies. The engineering plant and services incorporate the high degree of reliability essential to these studies. The laboratories and offices are not air conditioned. Chemical development laboratories and pilot plants are located on the adjoining production site.

The laboratories on the Research site overlook school premises to the north and the sports field, the River Lea and open country to the south. They are close to the town centre and easily accessible by road and rail. Obviously within the laboratories there are variations in the quality of the environment, but there is a continuous improvement programme. Some features of selected work environments are described below.

THE CHEMICAL RESEARCH LABORATORIES

A typical chemical research laboratory is used by six chemists including the Research Leader. Each chemist has a working bench, writing desk and a share in general laboratory equipment including three 2 m fume cupboards. Conditioned air equivalent to 12 changes an hour enters from a continuous grill set in the ceiling above the windows. An auxiliary air supply enters the laboratory from a plenum chamber directly above the face of the fume cupboards. All the air supplied to the laboratory is extracted through the fume cupboards at face velocity of 0.5 m/s, or better, and is ejected at high velocity through fans and ducts sited on the open roof of the building. The laboratories are quiet and well lit. The windows are double glazed, with a venetian blind between the two layers, and designed to separate and to swivel so that all the glass faces can be cleaned from inside the laboratory.

THE RADIOCHEMISTRY LABORATORY

The synthesis of chemicals specifically labelled with radioisotopes must be carried out within slot boxes and glove boxes which have good independent, air extraction. All isotopically labelled materials and apparatus used for synthetic work must be carefully controlled and monitored. The laboratory effluent, the laboratory workers and all surfaces must, in addition, be checked regularly to ensure absence of radioactivity. The laboratory environment where radioactive materials are used has been designed to provide a large margin of safety.

INSTRUMENT LABORATORIES

Electrical and electronic equipment may emit considerable heat. This heat is best removed from an instrument laboratory by installing cooling units which absorb the heat from the room and transfer it outside the building.

In a spectroscopic laboratory cleanliness, orderliness, organisation and space are essential. Facilities are needed for the preparation of solutions, the operation of the instruments and the storage of spectra records. As in many other laboratories, storage and retrieval of results derived from instruments is now achieved with microprocessors or computers.

CONTROL OF CHEMICALS

A chemistry department needs a large stock of chemicals and reagents. These must be stored in such a way that chemists know the materials that are available and where they are. Chemicals must be stored under conditions that take account of their properties; bulk solvents need to be kept in an external solvent store designed for the purpose; unpleasant and reactive chemicals are best stored on ventilated shelves in a room separate from the main collection; heat-labile or volatile materials require a cold room. The whole chemical stock needs to be listed and computers are valuable for this. A well run store avoids duplication of stocks, encourages the return of chemicals and allows old samples to be disposed of.

SPECIAL ENVIRONMENTS

Glaxo Operations (UK) Ltd at Ware make millions of metered dose aerosols containing drugs for the treatment of asthma. The propellants are mixtures of halons which boil below room temperature. Experiments in the formulation of these products must be carried out in a cold room which is fitted out as a laboratory. This is an example of a specialised working environment. Glaxo products are sold throughout the world. Storage tests are carried out under a variety of simulated climatic conditions in specific 'environment rooms'.

Supplies of a new pharmaceutical product intended for clinical trials must be made under conditions that are as good as those which will be used in the ultimate manufacturing process. Since the pharmacists who carry out this work may be exposed to biologically active dusts they wear either a dust mask or an *Airstream* helmet. The latter provides a dust-free stream of air — a micro-environment — over the face. Fluid-bed drying and tabletting are carried out in separate booths and care is taken to avoid cross-contamination between booths. The air supplied to and extracted from each booth is filtered. In this way the environment within the booth is controlled to satisfy the demands of a regulatory authority [1], and workers are protected within that environment.

CHEMICAL DEVELOPMENT PLANT

The development of a process by which a new chemical is made on a progressively increasing scale is carried out in a pilot plant. The design of a pilot plant used for a variety of processes is a specialised task drawing on the skills of architects, chemists, engineers and the manufacturers of plant and equipment. Two aspects of pilot plant design should be noted.

Since solvents are used in quantity, all the electrical equipment must be of a design approved as safe for use in atmospheres that might contain flammable materials. Local control of services is often achieved with pneumatic valves. Secondly, there is need for good general ventilation of the plant area to be supplemented by extraction of air from specific parts of the plant at which contaminants may concentrate. It may not be acceptable that these contaminants be extracted directly into the atmosphere, in which case the extracted air needs to be scrubbed free of contaminants before release.

SUMMARY

Some general improvements incorporated into the laboratories of Glaxo Group Research have been discussed. Special environments in chemical and pharmaceutical laboratories have also been described. There have been improvements in biology and pathology laboratories over the past few years and facilities for the care of animals have advanced dramatically.

Improvements will continue in response to various influences. Company policies and development of new projects linked with the advances of science and technology will all dictate changes in the working environment. If high quality staff are to be attracted, they will expect the best possible facilities for their work. Industrial legislation, and the demands of good laboratory practice [2,3] and good pharmaceutical manufacturing practice[1] which are enforced by quality assurance arrangements will ensure that, at least in the pharmaceutical industry, the quality of working environments will continue to improve.

REFERENCES

[1] *Guide to Good Manufacturing Practice,* Department of Health and Social Security, HMSO, London, 1971.
[2] *Non-clinical Laboratory Studies. Good Laboratory Practice Regulations,* Federal Register Volume 43, No. 247, p. 59986 Department of Health, Education and Welfare, Food and Drug Administration, USA, 1978.
[3] *Good Laboratory Practice in the Testing of Chemicals,* OECD, Paris, 1980.

Section 2:

EQUIPMENT

13

Options in the design of laboratory ventilation systems

A. G. Harris
Laboratory of the Government Chemist, London

INTRODUCTION

There are several ways in which laboratory ventilation can be carried out. This chapter reviews the various options available to the designer with particular reference to the systems adopted by the Laboratory of the Government Chemist (LGC). In addition, difficulties which arise in both design and workmanship will be considered.

The first question to be asked is, 'What is the purpose of laboratory ventilation?' The main requirement is to provide a safe working environment for the laboratory user; all other aspects, such as comfort for user, are of secondary importance. This need for safety must be emphasised particularly when energy saving schemes are considered.

Too often in new or refurbished laboratories, the ventilation system is relegated to a late stage in the design process. In the author's opinion this is a disastrous practice which can result in the system having to be incorporated in the design with difficulty, making it awkward to install, and often leading to the inclusion of examples of bad practice. Further, having been installed under difficult circumstances, it will doubtless be difficult to modify, maintain, clean and dispose of at the end of its useful life.

A properly designed laboratory ventilation system must have the capacity to cope with changes in internal conditions as well as the range of extreme environmental conditions likely to be experienced throughout the year.

TYPES OF EXTRACT SYSTEMS

Three main options are available for fume extraction systems

(1) The individual fan system.
(2) The main fan system.
(3) The dilution duct system.

(1) Individual fan system

In this system each fume cupboard or hood is run completely independently of its neighbours. Figure 13.1 shows such a system as a cross-section through a single storey laboratory block with a central corridor. Each fume cupboard or hood is provided with a condensate collector to prevent run-back from such operations as acid digestions. Centrifugal fans are recommended for this type of system since bifurcated fans are troublesome under corrosive conditions and have a resultant short life span.

The scheme illustrated has individual exhaust stacks, but it is common practice for these to be grouped together. The stack height will depend on a number of factors including the height of the surrounding buildings and the concentration of the effluent.

The make-up air plant should be designed to supply 90 per cent of the extracted air volume. This will ensure that each laboratory is maintained under negative pressure to prevent cross-contamination between neighbouring rooms. Make-up air must be introduced as far away from the fume cupboards as is possible, to ensure the minimum interference with their performance.

On the right-hand side of Fig. 13.1, an air intake is shown immediately below the exhaust stacks and close to the fans. This arrangement is undesirable and should be avoided as there is a strong possibility that it could lead to fume recirculation.

Designers of small laboratories frequently omit make-up air facilities as they are expensive and require, even in the simplest case, a heater battery to provide reasonable working and comfort conditions.

The omission of make-up air, however, is risky and all too frequently leads to the recirculation of fumes down the exhaust stack of a switched-off fume cupboard. The make-up air supply can be omitted if a large laboratory has only one or two fume cupboards. However, inevitably there will be problems when the laboratory is small compared with throughput of the extract system.

Individually switched exhaust fans have the apparent advantage of economic operation. However, they rely on the user to remember, or have sufficient concern for other workers, to switch them on. In large installations in which make-up air is used in conjunction with individually switched extracts there is a strong likelihood of recirculation or suck-back unless proportional flow control is provided in the make-up air system.

One disadvantage with individually switched systems is that they have to cope with high concentrations of fumes and cannot rely on any dilution brought about by the diversity of fume cupboard use.

(2) Main fan system

This system is the complete opposite to the individual fan system in that one large fan is used to extract from a battery of fume cupboards. Figure 13.2 shows a single storey laboratory block housing similar fume cupboards. In this

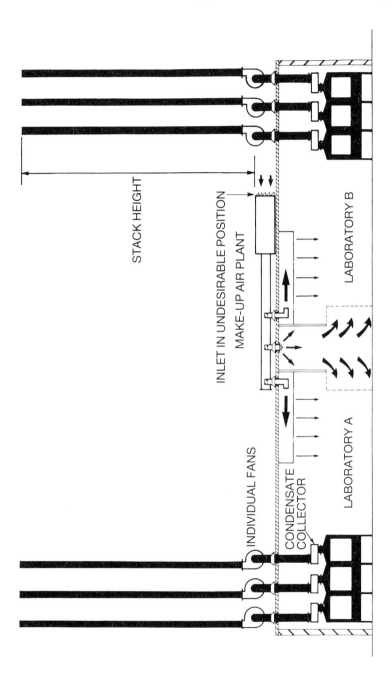

Fig. 13.1 — Individual fan system.

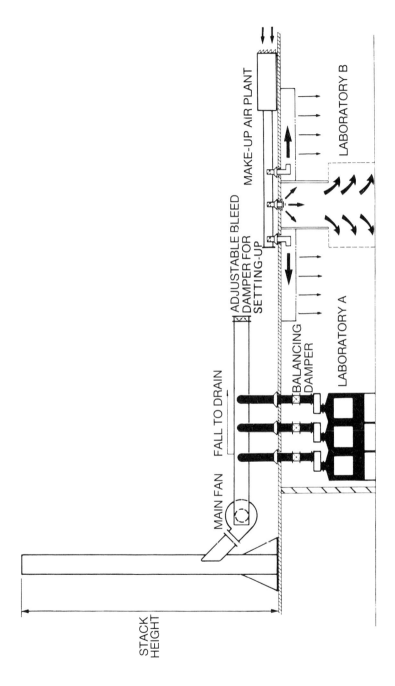

Fig. 13.2 — Main fan system.

system it is absolutely essential to use a centrifugal fan with a belt drive. This enables the duty of the fan to be adjusted to meet local conditions. A change-over motor is a necessity for stand-by purposes. Each duct branch must be fitted with a balancing damper. Experience at LGC has shown that the installation of an adjustable bleed damper at the end of the main duct helps considerably when balancing has to be carried out.

The same principles of siting and design of the exhaust stack apply as in the individual fan system and the provision of make-up air is essential. The air intake must be sited well away from influence of the exhaust stack and also away from exhaust fumes from vehicular traffic. It is important to arrange for the layout of the ductwork to fall to a condensate drain. Branch entries should be made into the upper side or top of the main duct to prevent condensate from running back.

The advantage of a main fan system is its simplicity. It is easier, more positive and has the assurance of a balanced air supply. Arrangements can be made to ensure that the fan is switched on before the staff arrive for work and switched off after closing time. The extract and make-up air systems should be interlocked, with the extract system as the dominant control. Where necessary the whole system can be left running to a timed or to a dual rate programme. Balancing this type of system is fairly easily achieved and fume dilution takes place within the main duct.

Disadvantages of the system are that fan failure will cause the shut-down of all the laboratories associated with it and, compared with the individually switched fan system, the energy consumption in fan power and the expulsion of heat in the extracted fume effluent is high. The possibility of extracting waste heat from the effluent stream by a heat exchanger or run around system has been considered by a number of users in the United Kingdom.

(3) Dilution duct system

The objective set for this type of system (Fig. 13.3) is to attempt to combine the advantages of the individual fan and main fan systems. Individually controlled fan exhausts are fed into a main duct which is in turn exhausted by a main fan of lower power than that used in the main fan system.

In order to maintain constant working conditions it is essential to admit the dilution air at the end of the duct most remote from the fan. A number of installers have failed to provide this feature with the result that fumes are recirculated into the laboratories through the exhausts of switched-off fume cupboards. Case investigations have shown that incorrectly specified, individual fume cupboard extract fans have been over driven by the main fan causing them to burn out. This results in the user ending up with, in effect, a main fan system.

The dilution duct system is the most expensive to install but it has been claimed to be more economical to run. It can be elaborated by day or night rate extract system control. As in the individual fan system, air balancing is more

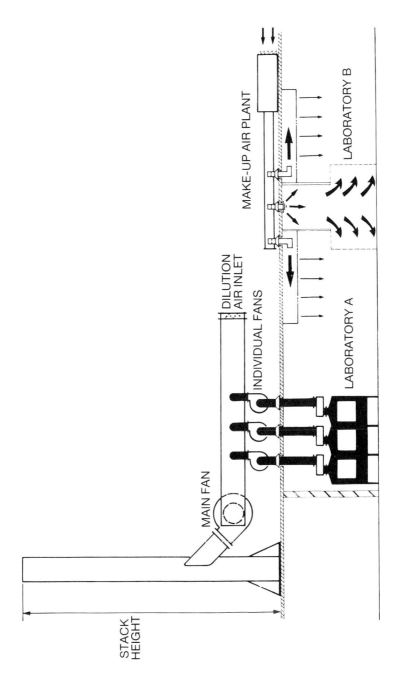

Fig. 13.3 – Dilution duct system.

difficult to achieve with the dilution duct system when compared with the main fan system.

Often there is conflict in the choice of system to be used. The cost factors of installation, operation, maintenance, and the facility for modification which will inevitably occur during the life of the system must be weighed carefully before the decision is made.

However, it must be emphasised that room air balance must be given first priority for safety reasons. If any arrangement is devised controlling the fume cupboard extract rate, whether an on/off system, or a graded control, a compensating control must be made on the supply air system.

Figure 13.4 shows a combination of features to be avoided in a system.

(a) Exhaust stack of inadequate height.
(b) Make-up air inlet too close to exhaust.
(c) Room air supply too close to fume cupboards.
(d) Exhaust fans too close to fume cupboards.
(e) No bleed damper has been included.

The layout of an installation in a small laboratory located in the Home Counties is shown in Fig. 13.5. The three fume cupboards each have individually controlled exhausts and no make-up air is supplied to the laboratory. The short stacks are terminated with cowls shaped like Chinese coolie hats. Bifurcated fans are used which are placed too close to fume cupboards and direct exhaust stacks have been installed with no facility for collection of condensate.

The users of the fume cupboards in this laboratory were subjected to a rain of condensate as they worked and fumes were deflected downwards from the coolie hat terminations into the laboratory windows. It was found also that unless all the fume cupboard exhaust fans were operating, fumes were recycled down the stacks of the switched-off cupboards. The problems were remedied firstly by connecting the three separate exhaust fan motors together on one control. Secondly, make-up air was provided. These modifications effectively produce an individual fan system.

DESIGN, MATERIALS AND WORKMANSHIP

A ventilation system must be thought through from the beginning, properly designed by experienced engineers and compromises avoided since they rarely work. An extract system is expensive to install and maintain and therefore demands good quality workmanship using materials that will withstand the working environment. Repairs are costly and disruptive to scientific staff. The extract system in LGC is based on glass-reinforced unplasticised PVC. The design of a system must eliminate cracks, crevices and points liable to collect effluent if a long and safe working life is to be achieved. It is advisable also to

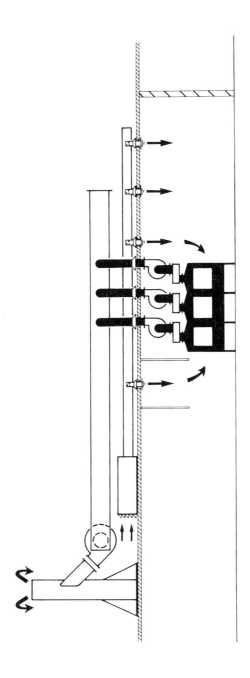

Fig. 13.4 – Common design mistakes.

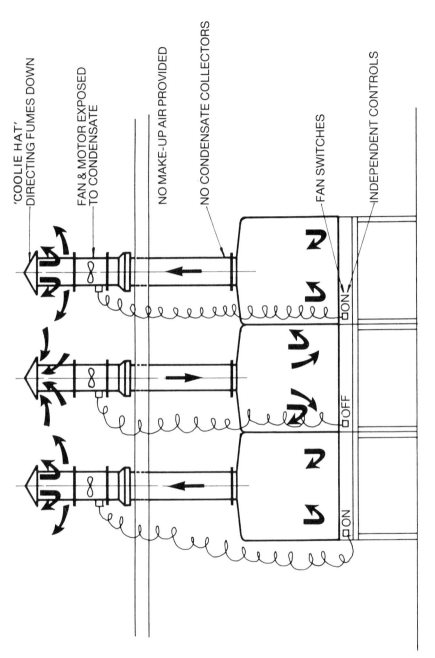

Fig. 13.5 — A design for fume recirculation!

furnish plenum supply ducts with cleaning ports as it will be necessary to clean the duct work after several years running even with commercial intake filters.

FUTURE DEVELOPMENTS

Work by a British Standards Institution committee on fume cupboards has resulted in proposals for the standardisation of exhaust rates[1]. This will lead possibly to standardisation of duct sizes. Once this has been achieved, it will be practicable to produce and use standard ductwork components thereby reducing both manufacturing and installation costs. This philosophy is being pursued by LGC and Fig. 13.6 shows a diagrammatic arrangement of a two fume cupboard extract system built from standard components.

Whatever developments occur in ventilation systems it is unlikely that there will be any change in the simple tenet — A user only gets the scheme he is prepared to pay for!

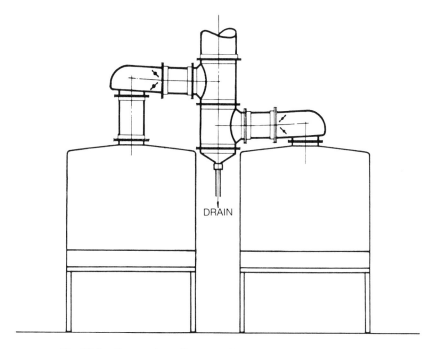

DRAIN

Fig. 13.6 — Ductwork configuration for a two fume cupboard system.

REFERENCE

[1] DD80: 1982, *Laboratory Fume Cupboards,* British Standards Institution, London, 1982.

14

Ventilating the laboratory without adversely affecting the performance of fume cupboards

P. J. Jackman

Building Services Research and Information Association, Bracknell

INTRODUCTION

The performance of fume cupboards can be markedly affected by the way the required supply air is introduced into the laboratory in which they are located. Difficulties are often encountered because of the considerable quantities of 'make-up' air usually required and the primary need to avoid air flow disturbances at the faces of the fume cupboards. Other practical requirements are also important, such as the prevention of draughts which could cause discomfort and impair laboratory processes.

This chapter describes an experimental study at the Building Services Research and Information Association (BSRIA) of some alternative arrangements for supplying air through a suspended ceiling to a laboratory, used for chemical synthesis, which incorporated purpose-designed fume cupboards. The study was conducted in a full scale functional mock-up of a section of the laboratory under design consideration.

The facility was also used to examine other features beyond the scope of this chapter. These included the arrangement of the air supply to the ceiling plenum, the lighting intensity at working level, and the performance of the fume cupboards under a wide range of test and simulated operating conditions.

TEST FACILITY

A full-scale replica of a section of a chemical synthesis laboratory was constructed in the BSRIA Laboratory. The overall dimensions of the mock-up were 8 m by 6 m by 3.3 m high. Along one of the long sides, three fume cupboards were simulated. The central cupboard was modelled in detail and was used for performance assessment purposes. The other two were elementary but functional simulations used, in conjunction with the central cupboard, to obtain the required extract flow rate and to enable the effects of air flows into adjacent cupboards to be studied. The layout of the test facility is shown in Fig. 14.1.

A proprietary suspended ceiling system was installed at a height of 2.7 m above floor level. This ceiling incorporated recessed light fittings and ventilation slots as illustrated in Fig. 14.2. By using damper strips fitted above each row of slots, air flow through sections of the ceiling could be reduced or prevented. The void above the ceiling was used as a plenum into which air from outdoors was supplied. The air supply system incorporated a centrifugal fan, a flow rate measuring device and control damper and an electrical heater battery by which the incoming air was heated. Its temperature was controlled at a predetermined value to within ± 0.2°C.

Two exhaust systems were used, one for the central fume cupboard and the other for both the adjacent cupboards. Provision was made for direct flow rate measurement in the central cupboard system, but to set up the other exhaust system, face velocity measurements were used as the flow rate criteria.

The locations of the central exhaust system discharge and the supply system fresh air intake were carefully chosen (Fig. 14.3) to minimise the re-entry of the tracer gas to be used in the assessment of fume cupboard performance. The intake was windward of the exhaust air outlets at the prevailing wind direction.

The mock-up laboratory was equipped with representative furniture. Two central benches and three carrels were included as indicated in Fig. 14.1.

ASSESSMENT OF THE MOST SUITABLE CEILING CONFIGURATION

A preliminary series of tests, mainly involving observations of smoke movement and subjective assessments, was made to determine the most suitable ceiling arrangement.

The three main ceiling slot configurations used in these tests were:

(a) All slots fully open,
(b) All slots within 1.6 m of fume cupboards closed, the remainder fully open, and
(c) All slots within 0.6 m of fume cupboards closed, the remainder fully open.

Tests were conducted with the air discharge from the ceiling slots:

(1) Vertically downwards.
(2) Horizontally in opposite directions.
(3) Part vertically and part horizontally.

The test conditions were:

Total supply air flow rate:	2.1 m³/s
Fume cupboard extract flow rate:	0.66 m³/s each (equivalent to a mean face velocity of 0.5 m/s with a 600 mm sash opening)
Supply air temperature:	21°C

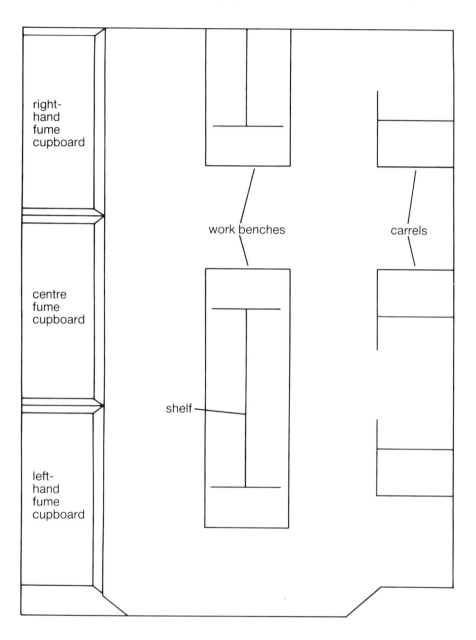

Scale 1:40

Fig. 14.1 – Plan view of laboratory mock-up.

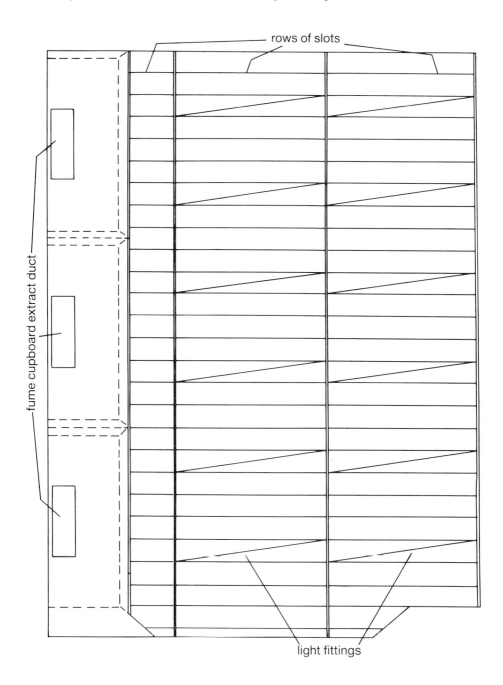

Fig. 14.2 – Layout of suspended ceiling.

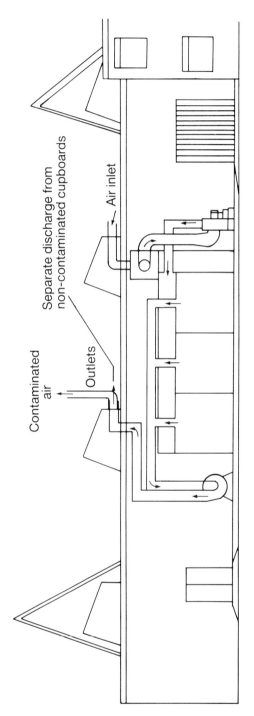

Fig. 14.3 — Vertical section of BSRIA laboratory showing position of test facility.

(1) Downward air discharge

The first observations were made with the ceiling slots set to discharge air vertically downwards. The supply air flow rate was equivalent to 16.8 l/s per metre of slot. Under this condition there was considerable air motion throughout the laboratory area with unacceptably high velocities, about 0.7 m/s, particularly at head height. The main pattern of air movement is shown in Fig. 14.4.

The recently published British Standards Institution Draft for Development on fume cupboards[1] indicates that the minimum distance from a supply grille to a fume cupboard should be 1.5 m and then only if the direction of discharge is away from the cupboard. To assess the effect of conforming to this distance requirement, observations were made with ceiling configuration (b) i.e. all slots within 1.6 m of the fume cupboard closed. This resulted in a strong air recirculation in a vertical plane near the cupboards, with upward flow next to the cupboard faces (Fig. 14.5). This rapid upward movement of air was observed to interfere with the air flow into the cupboard and was therefore considered unacceptable.

Reducing the closed section to 0.6 m (configuration (c)) resulted in a less marked recirculation pattern within the laboratory space, but some disturbing upward flow near the cupboards persisted (Fig. 14.6).

(2) Horizontal air discharge

Deflecting plates were fitted to the ceiling slots to produce a horizontal discharge of air in opposite directions. It became apparent immediately that this arrangement produced very disturbed air motion around the recessed light fittings and so flat light diffusers were placed across the base of the fittings, flush with the ceiling surface. The result was a discrete series of downward moving airstreams except that instead of occurring immediately below the slots, they were located centrally between the rows of slots where the two horizontal airstreams from adjacent sources met. The insert on Fig. 14.7 shows the pattern of air movement close to the ceiling and it may be compared with that for the previous condition shown on Fig. 14.4.

The maximum air velocity (0.38 m/s) at head level with all the ceiling slots open was less severe than previously but still considered too high for comfort particularly when related to a reduced supply air temperature. Observations made with the other two ceiling configurations revealed similar results as before but the general rate of air movement was less rapid.

(3) Combined vertical and horizontal air discharge

To generate a more uniform downward movement of air and thereby further reduce the room air velocities, slots were punched in the deflecting plates to produce a combination of downward and horizontal discharge (Fig. 14.8). The uniformity of flow was found to improve further with all of the main ceiling slots half closed.

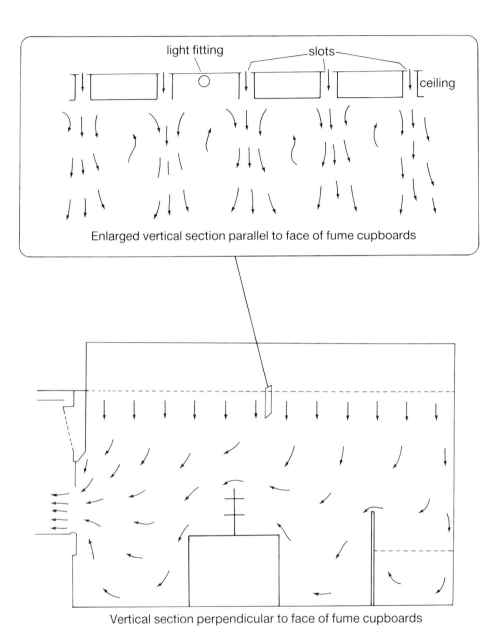

Fig. 14.4 – Diagram of air movement with downward air discharge and all ceiling slots open.

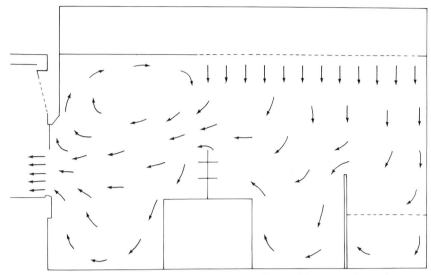

Vertical section perpendicular to face of fume cupboards

Fig. 14.5 – Diagram of air movement with downward air discharge and all slots
within 1.6 m of fume cupboards closed, remainder open.

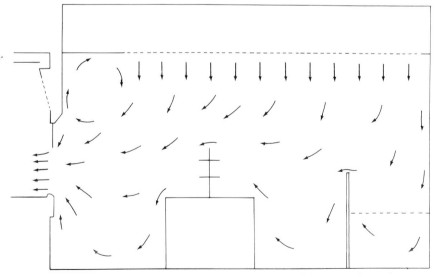

Vertical section perpendicular to face of fume cupboards

Fig. 14.6 – Diagram of air movement with downward air discharge and all slots
within 0.6 m of fume cupboard closed, remainder open.

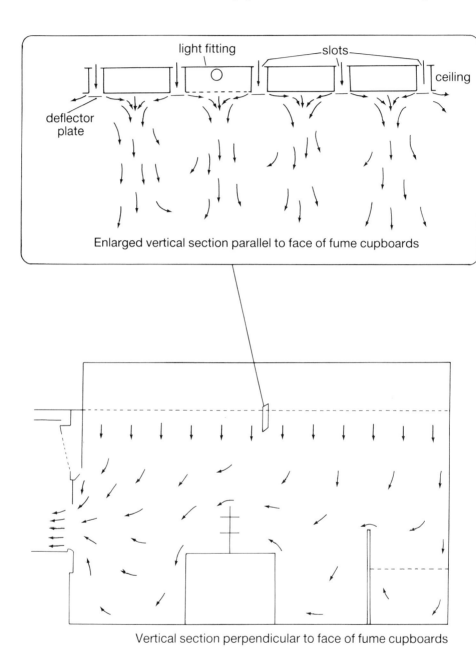

Fig. 14.7 – Diagram of air movement with horizontal air discharge and all slots open.

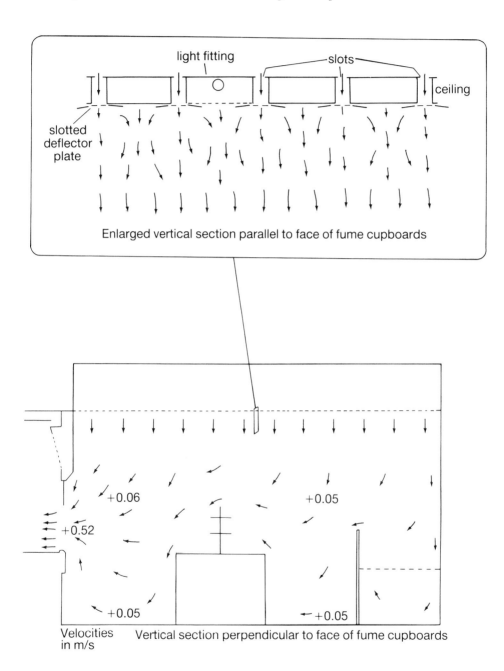

light fitting

slots

ceiling

slotted
deflector
plate

Enlarged vertical section parallel to face of fume cupboards

+0.06

+0.05

+0.52

+0.05

+0.05

Velocities
in m/s

Vertical section perpendicular to face of fume cupboards

Fig. 14.8 – Diagram of air movement with combined vertical and horizontal air
discharge and all slots open.

General observations confirmed that this ceiling configuration produced acceptable conditions and this was endorsed by more detailed room air velocity measurements. Generally, the velocities throughout the 'occupied' zone of laboratory were below 0.15 m/s although occasional excursions to peaks of 0.25 m/s or more were noted at some locations. Figure 14.8 includes a record of the average velocities at the positions indicated.

Some additional tests were conducted with varying supply air temperatures, the incoming air being as much as 6°C below the room air temperature. The only noticeable effect was a slight increase in the room air velocities near floor level.

FUME CUPBOARD PERFORMANCE

With the two outer cupboards operating, measurements of face velocities were made at the central fume cupboard with various sash openings. The results were satisfactorily uniform (i.e. within ± 20 per cent of the mean face velocity at the maximum working opening) and similar to those of previous tests on the same fume cupboard on its own in ideal conditions. It was thus shown that the laboratory air supply was not adversely affecting the air flow into the cupboard, and that the influence of the adjacent fume cupboards was not detrimental.

In addition, the containment performance of the central fume cupboard was measured using a tracer gas technique. These tests also demonstrated that acceptable performance was achieved but that it was necessary to utilise the whole ceiling area for introducing the supply air. There was significant loss of tracer gas containment when the area of ceiling close to the cupboards was blanked off.

The opportunity was also taken to examine the effects of other factors on containment performance, including the movement of a person both past the face of the cupboard and perpendicular to it.

SUMMARY

In this application, the use of a ventilated ceiling has been demonstrated to be an effective method of introducing air into the laboratory without adversely affecting the performance of fume cupboards. However, it was found necessary to produce as uniform a distribution as possible, and to introduce the supply air over the whole ceiling area rather than only through those parts remote from the fume cupboards.

ACKNOWLEDGEMENT

This investigation was undertaken by BRSIA under contract to Hoare Lea and Partners on behalf of the Boots Company Ltd. The author expresses his appreciation to these organisations for their kind permission to publish.

REFERENCE

[1] DD80:1982, *Laboratory Fume Cupboards*, British Standards Institution, London, 1982.

15

The dispersion of fume cupboard emissions in the atmosphere

A. G. Robins
Central Electricity Research Laboratories,
Central Electricity Generating Board, Leatherhead, Surrey

INTRODUCTION

From the point of view of its effects on ambient air quality, fume cupboard operation poses some particularly complex problems in addition to those associated with the aerodynamics of building affected dispersion. These extra problems result from the generally variable and uncontrolled nature of cupboard use coupled with the very wide range of possible emissions and corresponding threshold levels for health and related effects. A simple example serves to illustrate that these matters deserve serious attention.

Numerous studies have shown that the mean concentration level, as averaged over a period of about ten minutes or more, immediately downwind of a cuboid building, resulting from emissions from the walls, is given by:

$$CUA/Q \simeq 2\dagger$$

Given that a limit concentration (C_L) exists for the pollutant it follows that the maximum acceptable emission rate is:

$$Q \simeq 0.5 \, C_L UA$$

As an illustration, with $A = 100 \text{ m}^2$, $U = 2$ m/s and C_L (say, for health protection) $= 0.2$ mg/m^3, the limit emission rate is 20 mg/s (1.2 g/min). This emission rate, for a not particularly low C_L, is not at all large by practical standards.

For odours, the need to minimise likely problems probably depends on the nature and intensity of complaints. However, the topic is a difficult one because, although the olfactory system is very sensitive to changes in concentration level, it can become accustomed to maintained levels and, additionally, there are

---

† For explanation of symbols used in this chapter, see Appendix pp. 191–192.

considerable sensitivity variations between individuals. Difficulties are compounded because detection of odours is a function of existing background and, even under laboratory conditions, experimentally determined thresholds are found to be very variable. For health protection, however, there are published Threshold Limit Values (TLV)[1] and, at least from the designer's viewpoint, the situation is much more straightforward. Of course, other types of limits have sometimes to be taken into account; for example, for flammability, effects on materials, and effects on vegetation. As observed concentration levels are a function of the averaging time used in their measurement, it must be made clear at the outset that concern is generally with levels averaged over periods of, at least, five to ten minutes. An obvious exception is in the treatment of odours, where short period fluctuations have to be taken into account.

FLOW AND DISPERSION

The flow field
If a designer was concerned with a detailed analysis of the flow field and the dispersion of emitted material, then the effects of approach flow conditions, in particular atmospheric stability (density stratification), would have to be considered at some length. However, for the purpose of this chapter, such sophistication is unwarranted and analysis will proceed under the not unreasonable assumption that, near to a building, the dispersion of emissions from short stacks, above or adjacent to the building, is chiefly controlled by the effects of the building on the air flow, and is relatively insensitive to the details of the approaching atmospheric flow. Further downwind, or for taller stacks, the latter may dominate, but, in the present case, these matters may be largely neglected. A full account of this subject is given in an excellent book by Pasquill and Smith[2]. The chief interest in this account is in dispersion in the lower part of the atmospheric boundary layer, within which the wind speed increases with height and the turbulence decays. The depth of the boundary layer depends on the prevailing meteorological conditions; during fine summer weather the depth may be about 2 km, in windy weather about 700 m, and on clear nights with light winds no more than about 100 m. The flow field is, in general, continually developing, particularly over periods of a couple of hours; genuinely calm conditions are very rare.

Ahead of a building the flow is retarded and deflected to pass above and around the obstacle and there is a region of reversed flow, perhaps extending to a building height or so up wind and being about half that in depth. The general features of the whole flow are illustrated in Fig. 15.1; the details obviously depend on the geometry of the building and its environment. The upstream reversed flow, or recirculating flow, region is associated with the so-called 'horse-shoe' vortex which wraps itself around the building. Immediately downwind is the main recirculation region which stretches a distance, L_R, downwind

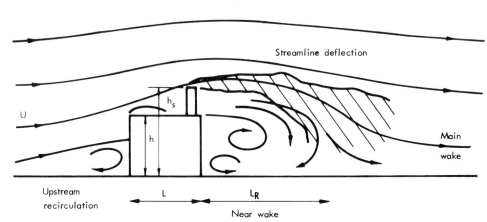

Fig. 15.1 – Flow field around a cuboid building.

of the building and has a maximum height equal to or somewhat greater than the building height; its maximum width is again somewhat greater than the building cross-flow width. Within this region mixing is generally very efficient and there is an exchange of air with the external flow due both to turbulent and mean-flow advections. For convenience, this downstream recirculation region will be termed the near-wake in order to differentiate it from the main-wake which follows it. In the main-wake, mean flow speeds are less and turbulence levels higher than in the approach flow, due mainly to the drag forces on the building and the turbulence generated in its vicinity. These differences decay with increasing downwind distance and eventually the flow returns, more-or-less, to its undisturbed form. Higher in the boundary layer the building's presence is mainly revealed by deflection of the mean stream lines, perhaps by substantial amounts.

The foregoing description is, at best, purely illustrative and, as implied, considerable variations may be brought about by building geometry, surrounding buildings and general features of site location. For virtually all cases, effects will also be sensitive to wind direction. A case in point being the flow over flat-roofed buildings, for when the wind direction is more than 15° off-normal to the front face, strong coherent vortex motions form above the roof due to the rolling up of the flow up the front faces. These motions produce strong downward streamline deflections above the roof and just downwind of the building. For air pollution considerations there is usually a worse wind direction and, in many cases, this is the one giving a near diagonal orientation.

General features of dispersion
Material emitted into the near-wake, from sources at or below roof level, is thoroughly mixed within the region and transported across the region's boundary both by turbulence and mean flow effects. This gives rise to a broad, deep,

ground-based plume of material which is carried away in the main-wake. In the near-wake, concentration levels are often claimed to be approximately uniform. This is a useful assumption for analytical purposes but must be used in cognisance of the variations which in reality exist, especially close to point sources, due to the finite mixing rates. Plumes from short stacks above roof level tend to remain quite coherent until, due to spreading and mean streamline deflection, they make contact with the boundary of the near-wake. Considerable entrainment takes place across the boundary which can produce quite high near-wake concentration levels, though less than for internal sources. However, a part of the plume remains aloft, although its height may be greatly reduced, and this may lead to a growth in ground level concentration just downwind of the near-wake. As the stack is made taller the concentration in the near-wake decays quite rapidly and the maximum ground level concentration is observed to occur somewhere downstream of the building. In these circumstances the emission behaves as if it originated from a source of less height than the actual stack. This height is often termed the 'effective' chimney height, and is less than the physical height because of the effects of mean streamline deflections and the comparatively high mixing rates in the wake.

The general behaviour of the concentration field with increasing stack height, for a cuboid-shaped building, is illustrated in Fig. 15.2[3]. In the Figure, as in the earlier illustrative evaluation, concentration levels are expressed in a non-dimensional form as CUA/Q. Non-dimensional forms of this type arise quite naturally from consideration of the conservation of mass or volume flux. A few points are worthy of note at this stage. The concentration level is proportional to the emission rate of the pollutant; in fact, the emission rate from all sources affecting the receptor. Concentration levels are inversely proportional to wind speed. This raises the question as to what speed should be used in the analysis of pollution levels. In the United Kingdom the annual average wind speed varies between about 4 and 7 m/s, at a height of 10 m above ground, being greatest in coastal areas and least in central England. Roughly speaking, there is a 5 per cent probability of exceeding twice this mean and a 10 to 15 per cent probability of speeds less than 2 m/s. In sheltered areas, so-called calm conditions may occur for up to 5 per cent of the time though, in fact, there is usually some air movement, particularly above the surface-affected layer, or in hilly regions. Lastly, concentration levels are a function of building geometry, here as $1/A$, and, additionally, the arrangement and sizes of surrounding buildings, and the general features of the site in question.

Concentration field

Some detailed information concerning concentration levels near buildings, at least for a limited range of geometries, will illustrate concentration fields. Figure 15.3 returns to the theme of an isolated cuboid building, of height h, and illustrates the nature of the air exchanges betweeen the near-wake and the external flow. An

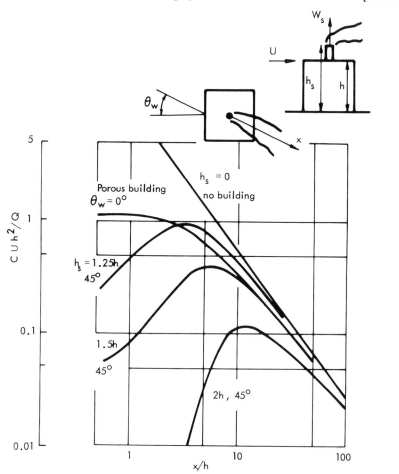

Fig. 15.2 – Effect of stack height on ground level concentration for a cuboid building when $W_s \simeq U$.

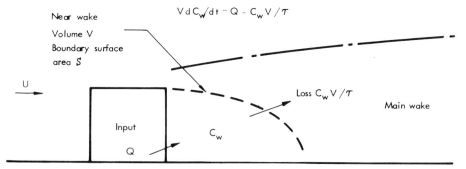

Fig. 15.3 – Processes in the near-wake region of a cuboid building.

analysis of this problem leads to the shown rate equation for the near-wake concentration, in terms of the region's volume, the pollutant emission rate, and the decay time constant. In the steady state:

$$C_w{}^* = C_w UA/Q = (U\tau/h)/(V/Ah) \tag{15.1}$$

Extensive wind tunnel measurements on block-shaped buildings by Fackrell and Pearce[4] lead to the following empirical relationships:

$$U\tau/h = 11.0(b/h)^{1.5}/[1.0 + 0.6\,(b/h)^{1.5}] \tag{15.2}$$

$$V/Ah = [1.75 + 1/(b/h)]\,L_R/h \tag{15.3}$$

$$L_R/h = 1.8(b/h)/[(L/h)^{0.3}(1.0 + 0.24\,b/h)] \tag{15.4}$$

In addition to permitting evaluation of (15.1), the length of the near-wake region is given by (15.4). Applicability of the above is probably restricted to:

$$1/3 \leqslant b/h \leqslant 3$$

Table 15.1 gives some evaluations based on the above, including the near-wake concentration for a unit emission rate. The volume given by (15.3) is normally somewhat greater than might be expected. It seems likely that the actual mean concentration in the whole near-wake volume may be about twice the value given by (15.1) – (15.4) and hence the exchange rates, V/τ, are a factor of two too large. The constants appearing in (15.3) were selected to give reasonable predictions of the ground level concentration field, away from the immediate vicinity of point sources. rather than the volume average.

Table 15.1

Properties of near-wake pollution for building wall emissions

h (m)	b (m)	L (m)	U (m/s)	L_R (m)	τ (s)	V/τ (m³/s)	C^*_w	C_w (mg/m³ for $Q = 1$ kg/s)
10	20	20	5	20	23	390	2.6	2600
20	20	20	2	29	69	460	1.7	2200
20	20	20	5	29	28	1200	1.7	860
20	20	20	10	29	14	2300	1.7	430
40	20	20	5	40	26	4600	0.87	220

The above model is a reasonable treatment of wall and roof emissions not possessing significant momentum or buoyancy. For roof emissions there will generally be wind directions leading to lower concentration levels than predicted by the above; by and large the range of such directions is narrow. Near-wake concentrations rapidly decay with increasing stack height, as is shown in Fig. 15.4. The almost exponential decay of concentration with stack height for $h_s > h$ should be noted; likewise the scatter in observed values of concentration. For stacks of sufficient height (say $h_s > 1.25$ to $1.5\ h$) much more rapid decay would be expected in atmospheric conditions producing low turbulence levels.

Of course, as has already been suggested, for a tall enough stack the maximum ground level concentration no longer occurs in the near-wake, but somewhere further down wind. Dispersion in these circumstances may be very crudely

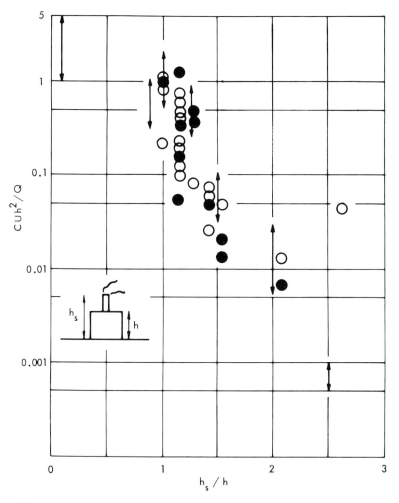

Fig. 15.4 — Near-wake concentrations of cuboid buildings. Key: $\updownarrow$: $L=b=h$; $\circ\ \bullet$: $L=b=1.5h$; $\circ$: stable atmosphere; $\bullet$: unstable atmosphere [5].

analysed in terms of effective chimney height, that is the height of a free-standing chimney the use of which leads to the same maximum concentration on the surface. The variations of effective height with actual height have been presented for a range of simple building shapes[6]; the data are reproduced in Fig. 15.5. As might be anticipated, no simple general relation emerges, though it is obviously true to say that tall narrow buildings have far less effect on effective height than short broad ones. Winds roughly aligned with roof diagonals represent the worst cases and, in these simple examples, would be assumed for the purpose of analysis.

Of course, effective stack height may be increased by the buoyancy and momentum contained in the emission. If the initial aim is to avoid significant ground level pollution in the near-wake, or indeed somewhat further down wind,

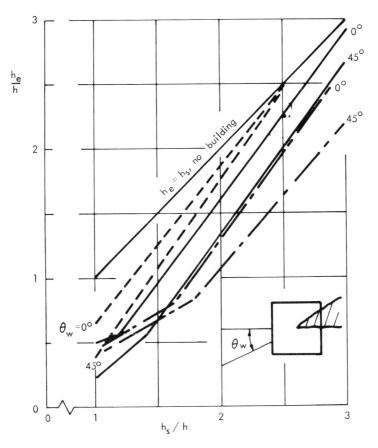

Fig. 15.5 – Effective stack height variations for block buildings with square roofs.
Key: side = $h/3$ – – –; side = h ———; side = $3h$ – —— .

then emission momentum is likely to be the main agent for increasing effective height. Additionally, sufficient exit speed should be provided to eliminate plume downwash in the wake of the stack. The results of some wind tunnel observations for a building with $b = L = 2h$ are summarised in Fig. 15.6. This has been plotted to show the physical character of the emission as a function of stack height and the square-root of the momentum flux, expressed as $(W_s/U)(d_s/h)$. The four characteristic plume forms are also shown. Clearly, to operate in regimes E or W, without the use of an exceptionally tall stack, requires

$$W_s \, d_s / Uh > 0.2 \qquad\qquad (15.5)$$

but great caution must be exercised in applying this result to buildings radically

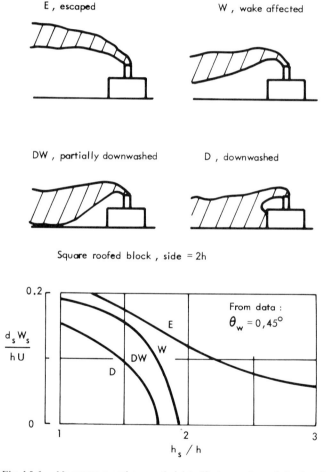

Fig. 15.6 – Momentum and source height effects on plume behaviour[7].

different from the one tested. For a fixed volume flux, (15.5) may be rewritten as:

$$Q_v/Udh > 0.16 \qquad (15.6)$$

Typically, these conditions require the emission speed to be about twice the wind speed. Based on the earlier summary of wind speed distributions, an emission speed of at least 10 m/s is indicated. In strong winds the plume will revert to behaving as in regimes DW and D, but due to dependence of the concentration field on $1/U$ this may well be of no great concern. It is worth noting that rain is most unlikely to enter a stack from which gases are being emitted at a speed of 10 m/s or more, and hence caps over the stack can be discarded. In practice, such caps simply ensure that what momentum there is in an emission is rendered valueless and that benefits due to plume buoyancy are greatly reduced.

The relative efficiencies of stack height and emission speed in the control of surface pollution, for a cuboid building, are illustrated in Fig. 15.7. So, to reduce

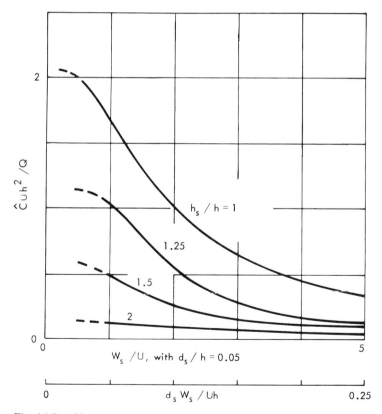

Fig. 15.7 − Variation of maximum surface concentration with stack height and efflux momentum, for a cuboid building. Cube data[3]: $\theta_w = 45°$; $d_s/h = 0.05$.

the consequences of emitting pollutant at a fixed rate at roof level with little momentum ($\hat{C}Uh^2/Q = 2$) by a factor of five requires a stack of height $1.5h$ and an emission speed of $1.3U$, or $1.25h$ and $2.5U$, or $1.0h$ and $4.3U$. As fan power is proportional to the cube of the emission speed, it will generally prove more economic to opt for a moderate stack height and an adequate minimum emission speed to ensure behaviour in the E and W regimes of Fig. 15.6. Estimates of plume rise, Δh, due to momentum may be made from the formula:

$$\Delta h = 3 d_s W_s/U \qquad (15.7)$$

where it has been assumed that there is little difference between the emission and ambient densities[8]. As (15.7) correctly indicates, the rise is a function of the emission momentum flux ($\pi d_s^2 W_s^2/4$) not the emission speed. Hence, for general purposes, the lower of the two 'x' scales of Fig. 15.7 should be used.

Calm conditions

Although buoyancy effects are generally likely to be insignificant, they may play an important role when winds are very light. Of course, momentum effects may also be of importance and the consideration of plume buoyancy is only relevant when vertical rise due to momentum is slight. Plume rise may be limited in the main for one of two reasons, stratification of the atmosphere, or negative emission buoyancy. In the former case, the final rise height varies as the 1/4 power of the initial buoyancy flux; for a comparatively small emission of 0.5 m³/s, with an initial temperature excess of 10°C, a final rise of about 30 m is likely[9]. It is interesting to note that for a given buoyancy flux, excessive emission speed may well result in a slight reduction in the final rise height. Negative buoyancy may exist in emissions from air-conditioned buildings during summer. If the temperature difference is large enough and the exit momentum small enough, the rise of the plume may be negligible and pollution will accumulate in low-lying areas[9]. Accumulations may also occur from horizontally directed emissions of low buoyancy. The value to which concentrations will rise depends on the volume of affected enclosed areas, the emission rate, and the duration of calm conditions. In the limit the source concentration may be reached, and for comparatively small volumes this may be attained in about an hour, or less. Clearly, ventilation intakes, etc. should not be sited in areas prone to such effects and it may be necessary to cease troublesome emissions when conditions are such that dispersion is very poor.

POLLUTION CONTROL

The approach to be described is summarised in Fig. 15.8.

Source characteristics

Although this is the first step in the procedure, it may be difficult to undertake with any great accuracy, and any uncertainties arising must be acknowledged

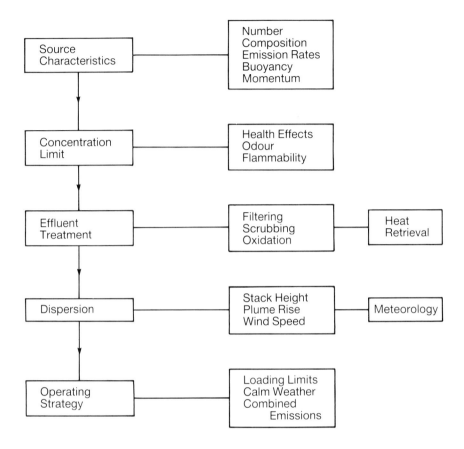

Fig. 15.8 – The approach to pollution control.

throughout the whole analysis. What is required is an assessment, for all relevant sources, of the range of substances involved, their emission rates (i.e. the values typically maintained over a period of several minutes or more) and the emission conditions (momentum and buoyancy). Included in this evaluation should be estimates of possible simultaneous emissions of different substances, taking into account all fume cupboards operable. If such an analysis is not possible, then worst condition figures may have to be substituted for the whole installation. As this is likely to overestimate actual conditions, perhaps considerably, there is a good incentive for deriving a reasonably accurate assessment.

Concentration limits
From the spectrum of emissions, the most problematical one, singly or in

combination, must be selected for design or assessment purposes. It follows from earlier that for a given source:

$$C_i = \alpha\, Q_i/UA$$

where the suffix denotes a specific substance and α is a function of stack height, etc. If the TLV is denoted by suffix T:

$$C_i/C_{Ti} = \alpha Q_i/C_{Ti} UA \tag{15.8}$$

Thus the emission of greatest concern is the one for which Q/C_T is largest. If α is approximately the same for all sources of interest then (15.8) can still be used, with Q as the total emission rate. The Health and Safety Executive (HSE) Guidance Note on TLVs[1] indicates how simultaneous emissions of several substances can be treated. If the actions of the constituents on health, etc. are known to be independent then (15.8) still applies, otherwise, for combined effects it is replaced by the index:

$$I = \sum_{i=1}^{n} C_i/C_{Ti} = UA \sum_{i=1}^{n} \alpha\, Q_i/C_{Ti} \tag{15.9}$$

If all α values are approximately the same then the index is a maximum, and the worst case identified when the sum of the individual Q/C_T values is greatest. If, due to variations in chimney height, etc., the values of α are significantly different, a more complex analysis based on the type of information given under 'Flow and Dispersion' (p. 169) may be needed.

Two kinds of threshold limits are often of interest, one related to adverse health or other effects of potentially harmful substances, and the second to odours. The TLV guide deals with health protection and, in setting maximum exposure levels for the population as a whole, a limit of a fortieth part of the time-weighted average (TWA) TLV would generally be involved (or a hundredth of the TLV for substances with a Ceiling Value), i.e. there is a limit concentration:

$$C_L = C_T/40 \ \ (\text{or } C_T/100) \tag{15.10}$$

The establishment of suitable limits for controlling odours is even more difficult than for toxic emissions. As outlined in the 'Introduction', it is the peaks or sudden changes in concentration which are readily detected and, for design purposes, these must be related to the mean concentration level so that emission limits can be evaluated. Observations of short-term concentration fluctuations made in wind tunnel studies of dispersion affected by buildings have shown results very similar to those obtained in plumes from ground level sources.

In summary, with $\hat{C}$ the local maximum ground level concentration and c' the standard deviation of the fluctuations:

$$c'/\hat{C} \simeq 1$$

Value exceeded 10% of time $\simeq 3.5\ c'$
Value exceeded   5% of time $\simeq 4.5\ c'$
Value exceeded   1% of time $\simeq 6.0\ c'$

Thus, in dealing with response to odours, it would be prudent to design to a mean concentration limit of about a tenth of the odour threshold:

$$C_L = C_T/10 \tag{15.11}$$

Values of the threshold limits themselves have been reported[10]. The above relations do not apply to elevated emissions, clear of building wakes, as in such plumes the value of c'/C is a strong function of the emission conditions[11].

Effluent treatment

Numerous methods are available for removing unwanted substances from fume cupboard air streams, the commonest being wet scrubbing, carbon filter adsorption, and thermal or catalytic oxidation. High removal efficiencies can be achieved with all methods when they are used in circumstances befitting their individual abilities. However, it seems unlikely that a single method can be recommended for general fume cupboard use. With filters there are problems associated with performance deterioration, due, for example, to high humidity or particle loading, detection of saturation, and avoidance of potentially dangerous mixtures in the filter bed.

The handling and treatment of saturated filters may not be straightforward, though the possibility exists of desorption of material stored in all filters servicing a large number of fume cupboards through a single, well designed, emission system. Wet scrubbing is, or course, restricted to use with soluble gases and leaves the problem of disposing of the used scrubbing fluid. Catalytic oxidation is unlikely to be of great value because poisoning of the catalyst greatly reduces its effective life. Thermal oxidation at a temperature of about 700°C may prove of use in disposing of odours and many hydrocarbon substances. However, oxidation of some materials can result in more troublesome emissions than the untreated gases, and partial incineration may make odorous emissions worse.

There may well be situations in which some form of treatment is a prerequisite to acceptable use of a fume cupboard installation; this is referred to later (p. 185). Obviously, careful consideration needs to be given to the selection of a suitable system and likewise to the control of the cupboard usage. In other cases it is a matter of comparing the capital and revenue costs, and operational limits, associated with effluent treatment with those of a properly designed stack system.

An associated area of concern is that of the heating or air-conditioning running costs required to compensate for fume cupboard use. For example, the heat loss rate due to the use of a single cupboard on a cold, winter's day may be of the order of 10 kW. However, in practical terms, it is very low grade heat and the prospects of exchanging large amounts between the exit flow and the compensating in-flow seem limited, particularly when the likely corrosive nature of the emission is considered. Obviously, keeping the through flow rate as low as is consistent with safe laboratory technique is important.

Dispersion

The physical processes have already been described under 'Flow and Dispersion'. It is instructive, however, to evaluate the benefits which can accrue by utilisation of stacks of various heights and this may be illustrated using the data shown in Figs. 15.4 and 15.5. In dealing with ground level concentrations likely to persist for 5 to 15 minutes, the following equation may be applied when the maximum concentration occurs downwind of the near wake:

$$\hat{C} = 0.2Q/Uh^2_e \qquad\qquad (15.12)$$

For some range of stack heights, the highest concentrations are found in the near-wake region, whereas for taller stacks they are found somewhere further downwind. Clearly, the amount of material which may be emitted to meet some ground level pollution limit increases as the stack height increases. The results obtained for the ratio Q/Q_h, derived from equations (15.1) to (15.4) and (15.12), together with the data in the Figures, are given in Table 15.2. As the

Table 15.2

Effect of stack height on permitted emission rate

$b/h, L/h$:	1/3	1	3
h_s/h		Q/Q_h	
1	1	1	1
1.1	2.4	1.7	1.5
1.2	4	2.0	1.7
1.4	8	2.6	1.8
1.6	12	6	2.6
1.8	18	10	4
2	24	16	6
2.5	47	36	13
3	70	60	24

data imply, $Q_h = Q$ when $h_s/h = 1$. A certain amount of personal judgement was used in preparing Table 15.2 and no great numerical accuracy is claimed. However, the picture revealed is a correct one.

The advantages of emissions possessing some momentum, or buoyancy, are found chiefly in light winds when, otherwise, ground level concentrations would be greatest. The data of Fig. 15.7 may be used to illustrate the point and these have been used to form the variations of maximum surface concentration with wind speed shown in Fig. 15.9. It is interesting to note that a doubling of the emission speed is roughly equivalent to increasing the stack height from h (a roof vent) to $1.25h$; thus, in this particular example an extra 5 m of stack achieves the same ends as an eightfold increase in fan power. Therefore, if all else permits, a moderate emission speed of 5 to 10 m/s with a tall enough stack should provide the required dispersion for the least revenue and capital expenditure.

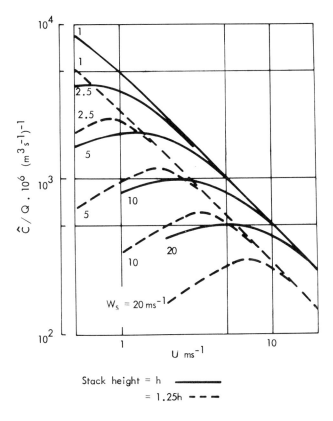

Fig. 15.9 – Variation of concentration with wind speed, for a cuboid building. Cube data: $\theta_w = 45°$; $h = 20$ m; $d_s = 0.5$ m. Key: stack height $= h$ ———, stack height $= 1.25h$ – – – –.

It must be realised that the arguments and analyses presented in this section are not particularly sophisticated, though they are probably adequate for the problem in hand. Concentration levels are affected not only by wind speed but by many other features of the atmosphere, so that a wide distribution of levels will be experienced down wind of a given source. It is likely that some of these values, during the course of the year, will exceed the limit levels set for design purposes. It may be argued that some allowance is, to a degree, included as the limit concentrations are small fractions of the equivalent thresholds; however, see p. 181 for the case of odours. For building wall or roof emissions the worst cases will occur in calm or near-calm conditions, when upward directed momentum, and buoyancy effects will be essential for pollution control. As noted earlier, such conditions may persist for up to 5 per cent of the time. For stack emissions the worst conditions are likely during sunny summer weather with light winds. Extra effective chimney height will somewhat ameliorate the situation but, because the frequency of occurrence of adverse conditions is likely to be less than 1 per cent, it may well be considered that the few higher than design limit concentrations are acceptable.

Operating strategy

It is clear from the foregoing that some form of operating strategy may well have to be adopted in certain circumstances. For a single fume cupboard installation it may be deemed prudent to:

(i)   limit the emission rates of substances, or combinations of substances, used; and

(ii) close down or further limit use, during adverse meteorological conditions.

For a multi-cupboard installation the above may still be required, plus control on simultaneous emissions from operating cupboards.

DESIGN IMPLICATIONS

General

Attention has been fixed largely on problems associated with isolated buildings, whereas interest often lies in building complexes. However, at least for estimation purposes, and perhaps for design, the methods and data enable air pollution consequences of fume cupboard operation to be evaluated. Additional, useful data is available [12–14]. In general, though, the possible approaches to a particular problem are, in order of increasing cost:

(i)   Analysis of experience with similar situations.
(ii) Extrapolation or interpolation of data from related situations.
(iii) Model tests.
(iv) Full-scale tests.

The cost, and duration, of each of these depends very much on the problem of interest and cannot therefore be stated meaningfully. As a guide, (i) may be an exceedingly brief procedure but probably rarely usable, (ii) may take anything from the order of a day to tens of days. Model tests involve the use of a wind tunnel, or similar facility, for a period usually measured in weeks, and field tests normally require much longer durations to cover the vagaries of atmospheric motions.

Nearby large buildings

If the emitting building lies well within the near-wake of an upstream building then the latter building will control the initial diffusion. Even when the two are far enough apart for this not to be so, significant effects can occur. These are strongest when the upstream building is diagonally oriented with respect to the oncoming wind. It has been found[15] that with a cube, oriented at 45° to the wind, upstream of an isolated stack significant effects occurred when the cube centre to stack spacing was less than about seven cube heights. The findings are summarised in Table 15.3 in terms of the ratio of the maximum concentrations downstream of the stack with and without the cube in position.

Table 15.3

Effect of an upstream cubic building
(height $= h_u$)

$\hat{C}/\hat{C}$ (no building):	2	3	4
h_s/h_u		Separation x/h_u	
0.75	6.7	3.7	2.4
1	6.7	2.5	1.2
1.25	4.7	1.1	*
1.5	2.7	*	*
1.75	1.0	*	*
2	*	*	*

* Effect not possible.

An emission from a stack unaffected by near-wake effects may be advected against a suitably tall down wind building. In this case, the maximum concentration experienced on the building will be roughly that which would have occurred in the plume at that position in the absence of the building. In the most adverse light wind conditions elevated plumes spread very slowly and dilution due to initial plume rise can be very important. A reasonable expression for the maximum concentration (averaged over a period of 5 to 15 minutes) within a plume in such conditions is:

$$C/Q = 1/2\pi U[\sigma_s^2 + (0.02x)^2]$$ (15.13)

If σ_s is small very large concentrations may occur within tens of metres of the source; these concentrations will not, of course, exceed the source concentration. For a plume which has risen 10 m or so, due to its initial momentum and buoyancy, σ_s may well be about 1 m and, at a wind speed of 2 m/s, C/Q will fall from 0.069 (s/m^3) at 20 m to 0.016 at 100 m. Clearly, problems can arise if the plume material is able to enter the building and should it be necessary to control concentration levels then very small emission rates could result. For example, if the 0.2 mg/m^3 limit discussed in the Introduction were applicable then the maximum emission rate would be 2.9 mg/s (0.17 g/min) for a 20 m separation, or 12.5 mg/s (0.75 g/min) for 100 m. In average conditions, the spreading constant, 0.02, would rise to about 0.1 and U would be about 5 m/s, and the emission limits would be 30 and 630 mg/s, respectively.

Ventilation intakes

Clearly, intakes should be sited such that the resulting concentration of effluent material within the building is satisfactorily controlled. Wilson [16,17] gives some information concerning building surface concentrations likely to result from emissions from the building. Of particular interest is the result that the maximum concentration likely to be experienced at a receptor, distance ρ, down wind of a flush source is given by:

$$CU\rho^2/Q \simeq 10$$ (15.14)

with the obvious limits that C cannot exceed the source concentration or, for some receptors, fall below the near-wake concentration. For elevated sources in the place of the flush vent, (15.14) would generalise to:

$$CU\rho^2/Q = \alpha(h_s)$$ (15.15)

However, the concentration in the building may well exceed the level directly evaluated from (15.14) or (15.15). Consider a building of volume v through which ventilation air is drawn at a rate of Q_v. Let the release rate of contaminant in a fume cupboard be Q and the resulting room concentration be C_R. At the vent position it follows from (15.15) that:

$$C_v = \alpha(Q + C_R Q_v)/U\rho^2$$

since Q is a mass emission rate and Q_v a volume flux. Equating the loss and gain of pollutant in the steady state gives:

$$Q_v C_v = Q_v C_R$$

so that:

$$C_R U \rho^2 / Q = \alpha / (1 - \alpha Q_v / U \rho^2) \qquad (15.16)$$

and, of course, the right-hand side of (15.16) is greater than α. In Table 15.4, C_R is evaluated from (15.16) for two values of $\alpha; \alpha = 10$ being for a flush vent and $\alpha = 1$ for a typical short stack. It is assumed that $Q_v = 1$ m³/s and $U = 2$ m/s.

Table 15.4

Concentrations within emitting buildings; $Q_v = 1$ m³/S, $U = 2$ m/S

α:	10	1
ρ(m)	$C_R U_\rho^2 / Q$	
2.5	50	1.09
3	23	1.06
4	15	1.03
5	13	1.02
10	11	1.01

The time constant for the growth of C_R following commencement of an emission is given by:

$$t_v = v / (1 - \alpha Q_v / U \rho^2) Q_v \qquad (15.17)$$

In the cases considered above t_v would vary from $5v/Q_v$ (for $\alpha = 10$ and $\rho = 2.5$ m) to v/Q_v. Thus it seems likely that only for prolonged emissions or small buildings will the levels indicated in Table 15.4 be attained. Anyway, the design target should be for a small enough α and a large enough ρ^2, so that C_R is likely to be little different from that expected directly from (15.15). However, the variation of α with stack height is likely to be very dependent on geometry and orientation. It is worth noting that if the intakes are in the building walls, in a region experiencing the near-wake concentration C_w, then an analysis similar to that resulting in (15.16) shows that $C_R = C_w$, calculated from the pollutant emission Q. In such a case proper control of C_w ensures proper control of C_R. Other ways of ensuring that the effective value of α is small include arranging ventilation intakes such that only a fraction of incoming flow is affected by the emission at any one time, or simply ensuring that ρ is large enough.

Examples of emission limits
Finally, it is interesting to evaluate some specific limits. For this purpose, the case illustrated in Fig. 15.9 concerning emissions from a centrally placed stack,

of diameter 0.5 m, in the roof of a 20 m high cubic building will be considered. In fact, limits have been evaluated for two stack heights, $h_s/h = 1$ and 1.25, and five emission speeds, 1, 2.5, 5, 10 and 20 m/s, and the results are given in Table 15.5. For the low emission speed cases the emissions have been calculated for a wind speed of 1 m/s and, therefore, some restriction on use might have to be applied in lighter winds. For the other cases, the maximum concentrations taken from Fig. 15.9 define the worst effects. It is pointless attempting to cover the whole range of likely emitted substances, and just four with quite low threshold values have been considered. However, there are many toxic substances with thresholds at the 0.1 mg/m^3 level, but very few with odour thresholds below that of hydrogen sulphide.

In general, elimination of likely odour problems implies lower emission limits than for attaining the required concentrations for toxic substances; there

Table 15.5

Emission limits for roof source and short stack on a cubic building; $h = 20$ m, $d_s = 0.5$ m

Emission speed (m/s)			1	2.5	5	10	20
Substance	Threshold (mg/m^3)	h_s/h		Maximum emission rate (mg/s)			
Health protection (TLV)							
Carbon disulphide	30	1	160	200	380	750	1500
		1.25	270	310	630	1200	2500
Hydrogen chloride	7	1	15	19	35	70	140
		1.25	25	29	58	120	230
Hydrogen sulphide	14	1	73	95	180	350	690
		1.25	130	150	290	570	1200
Sulphuric acid	1	1	5.2	6.8	13	25	49
		1.25	8.9	10	21	41	83
Odours							
Carbon disulphide	0.63	1	13	17	32	63	120
		1.25	23	26	53	103	210
Hydrogen chloride	14	1	290	380	700	1400	2700
		1.25	500	580	1200	2300	4700
Hydrogen sulphide	0.0066	1	0.14	0.18	0.33	0.66	1.3
		1.25	0.24	0.28	0.55	1.1	2.2

are exceptions of which hydrogen chloride is an example. In the configuration tested the use of a stack projecting 5 m above the roof is roughly equivalent to a doubling of the emission speed for the roof level vent. An eightfold increase in maximum emission compared with a roof vent with low emission speed (1 m/s in the Table) can be reached with the 5 m stack and a 10 m/s emission speed. Thus, in the case chosen, it does not prove too difficult to overcome the problems that would occur for a low momentum roof emission.

The quoted examples are for single substance emissions and the question arises as to how to design for multi-substance emissions. The HSE Guidance Note on TLVs[1] provides information here and, in the absence of any better method, it is suggested that the same techniques be applied for odorous emissions. If substances act independently, for example if they locally affect different organs of the body, then it is adequate to design according to the most potentially harmful component. Otherwise an index is formed, cf. equation (15.9), as:

$$I = \sum_{i=1}^{n} C_i/C_{Li}$$

and I should not exceed unity. As an example, it follows from Table 15.5 that as in a specific case the individual limits for sulphuric acid and hydrogen chloride are 41 and 120 mg/s, an emission rate of 20 mg/s of sulphuric acid together with 60 mg/s of hydrogen chloride would also be acceptable.

CONCLUSIONS

It would seem that in most cases relatively modest emission speeds and stack height can ensure adequate dispersion of fume cupboard emissions. Following the methods outlined, it should be possible to identify likely problems and formulate reasonable solutions. However, some situations may demand more specific treatment, such as wind tunnel modelling, to study the complex effects that can occur in practice. Probably the most difficult aspect of the problem is the determination of the likely range of emissions and emission rates, but without a realistic assessment in this area the remaining analysis is of little value.

If it is accepted that there are emission limits for any given fume cupboard installation then operating procedures will require modification. This will imply that cupboards emitting from a given building cannot, in general, be operated independently. Of course, the design may be chosen so that most operations can be performed without particularly severe restrictions. What is meant by 'most' is a decision for the user − it will certainly reflect both financial constraints and the inconvenience of dealing with the remaining cases. It might be quite acceptable to restrict use to times when meteorological conditions are favourable, or other usage is light.

ACKNOWLEDGEMENTS

Gratefully acknowledged are several helpful conversations with Mr R. Hotchkiss (CEGB, Marchwood Engineering Laboratories) and Dr D. J. Hall (Warren Spring Laboratory). Much of the work concerning dispersion affected by buildings has been undertaken in collaboration with Dr J. E. Fackrell (CEGB, Marchwood Engineering Laboratories) and Dr J. C. R. Hunt (Cambridge University). This work was carried out at the Marchwood Engineering Laboratories and is published by permission of the Central Electricity Generating Board.

REFERENCES

[1] *Threshold Limit Values for 1980,* Guidance Note EH 15/80, Health and Safety Executive, HMSO, London, 1981.

[2] F. Pasquill and F. B. Smith, *Atmospheric Diffusion,* 3rd edn, Ellis Horwood Ltd, Chichester, 1983.

[3] A. G. Robins and I. P. Castro, *Atmos. Environ.,* 1977, **11**, 291.

[4] J. E. Fackrell and J. E. Pearce, *Report RD/M/1179N81,* Central Electricity Generating Board, 1981.

[5] D. G. Smith, Influence of meteorological factors upon effluent concentrations on and near buildings with short stacks, *Proc. 68th Annual Meeting APCA,* Paper 26.2, Boston, USA, 1975.

[6] A. G. Robins and J. E. Fackrell, Laboratory Studies of dispersion near buildings, *Proc. Commission of the European Community Symposium on Radioactive Releases and their Dispersion in the Atmosphere following an Hypothetical Accident,* Vol. II, p. 971, Risø National Laboratory, Denmark, 1980.

[7] D. J. Koga and J. L. Way, Effects of stack height and position on dispersion of pollutants in building wakes, *Proc. 5th International Wind Engineering Conference* (Ed. J. E. Cermak), p. 1003, Pergamon Press, 1979.

[8] G. A. Briggs, Plume Rise, *USAEC Report No. TD-25075,* US Atomic Energy Commission, 1969.

[9] J. S. Turner, *Buoyancy Effects in Fluids,* Ch. 6, Cambridge University Press, Cambridge, 1973.

[10] G. Leonardos, D. Kendall and N. Barnard, *J. Air Poll. Control Assoc.,* 1969, **19**, 91.

[11] J. E. Fackrell and A. G. Robins, *Boundary Layer Met.,* 1982, **22**, 335.

[12] D. J. Wilson and D. D. J. Netterville, *Atmos. Environ.,* 1978, **12**, 1051.

[13] R. S. Thompson and J. G. Lombardi, *Dispersion of roof-top emissions from isolated buildings – A wind tunnel study,* USEPA-600/4-77-006, 1977.

[14] A. H. Huber, W. H. Snyder and R. E. Lawson, *The effects of a squat building on short stack effluents – A wind tunnel study,* USEPA-600/4-80-055, 1980.

[15] C. F. Barrett, D. J. Hall and A. C. Simmons, *Dispersion from chimneys downwind of cubical buildings – A wind tunnel study,* NATO/CCMS 9th International Meeting on Air Pollution Modelling and its Application, Toronto, August 1978.

[16] D. J. Wilson, *ASHRAE Trans.,* 1976, **82**, 1024.

[17] D. J. Wilson, *ASHRAE Trans.,* 1977, **83**, 157.

APPENDIX: NOTATION USED

A	frontal area of building
b	cross-stream dimension of building
C	mean concentration
C_L	limit concentration
C_R	room concentration
C_T	threshold concentration
C_v	ventilation inlet concentration
C_w	near-wake concentration
C^*_w	$C_w \, UA/Q$
$\hat{C}$	maximum concentration
c'	standard deviation of concentration fluctuations
d_s	source diameter
h	building height
h_e	effective stack height
h_s	source height
h_u	up wind building height
Δh	plume rise
I	index of combined effect of pollutants
L	streamwise dimension of building
L_R	near-wake (recirculation region) length
ρ	source–receptor separation
Q	source strength
Q_v	volume flux
Q_h	limit source strength for roof/wall sources
S	surface area of near-wake boundary
t	time
t_v	time constant associated with C_R change
U	mean wind speed affecting emission
V	near-wake volume
v	room volume
W_s	emission speed (upwards)
x	distance down wind of source
α	non-dimensional concentration

θ_w	wind direction relative to building
σ_s	effective source size
τ	time constant associated with C_w change

Suffix

i	specific substance

16

Laboratory fume cupboards and energy conservation

L. Thomas
Shell Research Ltd, Sittingbourne, Kent

INTRODUCTION

Laboratory designers have always recognised that fume cupboards provide the major element in the ventilation of a laboratory block and often, in individual laboratories, excess ventilation occurs because of the need to satisfy fume cupboards. In recent years a substantial increase in the use of fume cupboards has occurred because health and safety requirements have received increased attention as more has become known about the potentially harmful nature of many commonly used chemicals. This trend, which is continuing, often results in gross over-ventilation of laboratories and consequent increase of running costs. There is a paradox in this situation in that the increased use of fume cupboards with resultant reduction in work on open benches probably provides opportunities for lower rates of ventilation to be satisfactory. A laboratory of, say, ten years ago considered to be satisfactory with ten air changes per hour may now, with more fume cupboards installed, be satisfied with six, but the increased use of fume cupboards could easily have pushed the ventilation rate to 40 air changes per hour or more. The increased running cost has been referred to in the context of increasing ventilation rates but the situation has been compounded by the enormous increase in energy costs which occurred in the same period and which is continuing.

Over-ventilation, as well as causing increased running costs, also imposes capital cost requirements. Larger ventilation ducts, fans, heating plant, etc. have to support the excess air being handled. Designers have to take these factors into account and apply innovations to tackle them.

OPERATION OF CONVENTIONAL FUME CUPBOARDS

It is generally agreed that a face velocity at a fume cupboard (Fig. 16.1) of 0.5 m/s will prevent fumes from escaping from the chamber, and that a sash

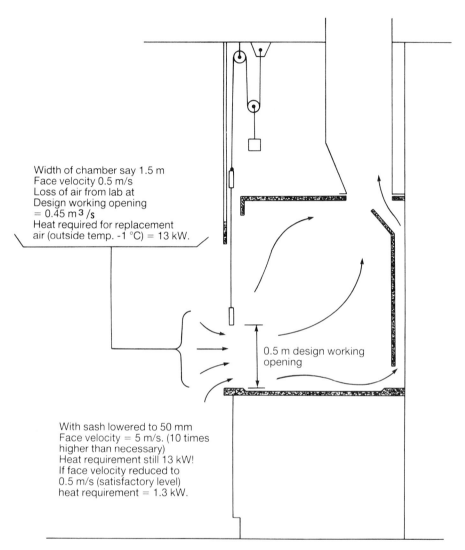

Width of chamber say 1.5 m
Face velocity 0.5 m/s
Loss of air from lab at
Design working opening
$= 0.45\ m^3/s$
Heat required for replacement
air (outside temp. -1 °C) = 13 kW.

0.5 m design working
opening

With sash lowered to 50 mm
Face velocity = 5 m/s. (10 times
higher than necessary)
Heat requirement still 13 kW!
If face velocity reduced to
0.5 m/s (satisfactory level)
heat requirement = 1.3 kW.

Fig. 16.1 – Conventional fume cupboard.

opened to 0.5 m will enable most operations to be carried out with the user's face in fresh air and protected by the sash from any accidents that occur in the chamber. These two figures, together with the width of the fume cupboard, are used to decide on the air quantity required to be discharged through the fume cupboard and also to be replaced by the fresh air system. For example, a 1.5 m wide fume cupboard will impose a theoretical demand of 0.375 m^3/s on both the extract and the fresh air systems. In practice 20 per cent excess will be allowed to compensate for probable lack of uniformity in air velocity across the exposed area under the sash, hence the systems will be designed for 0.45 m³/s. To raise this quantity of air from $-1°C$ to $19°C$ requires 13 kW.

The volume of air drawn through a fume cupboard is virtually constant regardless of the position of the sash, except when the sash is in its almost closed position. The velocity of the air entering the chamber, therefore, increases as the sash is lowered so that, at low positions, the velocity can be high enough to disturb experimental conditions. There is no reason connected with containing the generated fumes for the velocity of incoming air to be greater than 0.5 m/s, but with a conventional fume cupboard this occurs whenever the sash is lower than 0.5 m.

More recent conventional fume cupboards incorporate an air entry position at high level in the chamber so that the sash closes it completely at its uppermost position and opens it progressively as the sash is lowered. This is called by-pass air entry. If properly sized and located this secondary air inlet can ensure that the air enters the chamber below the sash at constant velocity. This is accomplished, however, by losing air from the laboratory through the by-pass and there is no reason connected with containment for this to be necessary.

In the example given earlier of a 1.5 m wide fume cupboard operating while the outside temperature is $-1°C$, 13 kW of heating is required for replacement air no matter where the sash is positioned, except when almost closed; this applies to both types of conventional fume cupboard described. For sash positions lower than 0.5 m this heating requirement is more than is needed for the fume cupboard's primary function, namely containing generated fumes.

ACHIEVED DESIGN CHANGE

An addition to a conventional fume cupboard has been developed and used in over 100 units during the last three years. This consists of a linkage system, operated by the sash balance weight, which operates a butterfly damper in the extract duct connected to the head of the fume cupboard (Fig. 16.2). The balance weight is supported on a double pulley system so that its stroke is half the sash's travel and an inconveniently large movement of the links is avoided. Slots and holes in the links allow adjustment of connecting pin locations to obtain a linkage geometry that will relate the movements of the sash and the damper properly.

A problem with the non-linear response of the butterfly damper is more apparent than real. Some variation in air velocity over the face area of a fume cupboard occurs with the best designs and experience has shown that the variation with the linkage in use is no worse than is found in conventional fume cupboards with the sash set at the design opening of 0.5 m. The imperfection in the response of the damper is perfectly acceptable against the problems of the conventional fume cupboards which the added device is intended to tackle.

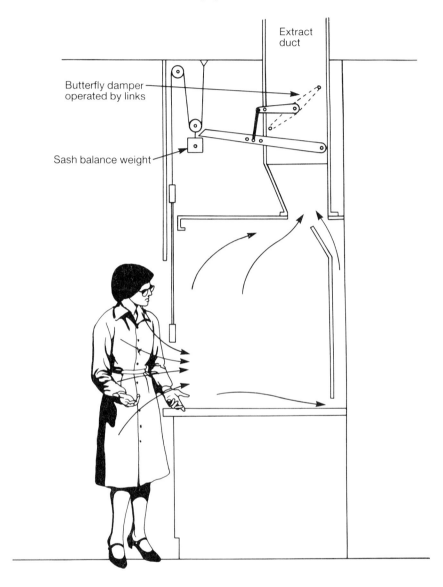

Fig. 16.2 – Conventional fume cupboard with added linkage and damper.

The linkage system requires no action from the user and, in fact, as Fig. 16.2 shows, is not directly accessible to the user. An energy demand of 13 kW arises under the weather condition described earlier when the sash is 0.5 m high in a 1.5 m wide fume cupboard and this remains if the sash is raised. Lowering of the sash, however, reduces the energy demand almost proportionally so that at a height of 50 mm it is only about 1.3 kW. Proper use of a fume cupboard should result in the sash being at its lowest position most of the time, so considerable energy economies arise with each fume cupboard. The same fact allows significant diversity of demand to be applied when ranges of fume cupboards are connected to a single extract system and served by a single fresh air system. In this way reductions in ducting and equipment sizes can be gained.

REPLACEMENT AIR

Addition of the device described converts a conventional fume cupboard to a variable volume unit and there will be imbalances between the fresh air supply and the extract system during normal operation unless action is taken to preserve a balance. More than one method is available, but a signal is required from the fume cupboard indicating the position of the sash and, therefore, the demand the unit is making on the fresh air system. A potentiometer fastened to the spindle of the damper is a convenient device for obtaining this signal. When more than one fume cupboard is supported by a single fresh air system the electrical outputs from each potentiometer can be summed to provide a signal proportional to the total instantaneous demand on the fresh air system.

An elegant method of achieving the variation of air volume in the fresh air system uses an electronic speed controller on the fan drive, automatically adjusted by the total signal from the potentiometers. Speed controllers are becoming widely available and economical. When a fan on a fresh air system serves more than one laboratory where fume cupboards are used, motorised butterfly dampers in the legs of the system serving the individual laboratories can provide variable air volumes. The motorised dampers can be adjusted using the total signals from the potentiometers in the laboratories.

HEAT RECLAMATION

The discharged fumes from ranges of fume cupboards are unlikely to be much warmer than 22°C. This is, however, sufficient to provide heat for incoming air to the fresh air systems when the outside temperature is 10°C or below. Run-around systems and plate heat exchangers are useful devices to transfer this low grade heat to the fresh air intake without risk of recirculation of the extracted fumes.

Run-around systems have been in use on laboratory extract plants at Shell's Sittingbourne Research Centre since early 1979 and excessive corrosion from fumes has not been evident. It is uneconomic to provide these devices for individual fume cupboard discharges; in fact, large extract plants serving several fume cupboards have to be provided.

SUMMARY

The energy consuming problems arising from fume cupboards can be tackled. The philosophy should be to allow the fume cupboards to extract from the laboratory only as much air as is needed for containment of fumes, and, having limited the amount of discharged air being handled, regain heat from it to help raise the temperature of the fresh replacement air.

17

Co-ordination of responsibilities for fume cupboard safety

A. L. Longworth
Formerly Imperial Chemical Industries plc, Alderley Edge, Cheshire

Until recently, fume cupboard performance has been specified in relation to categories of usage and particular air velocities at the face of the open sash, and by reference to a British Standard Specification[1]. Discriminating manufacturers and users have also used smoke visualisation of the airflow pattern to provide qualitative, but non-quantitative, indications of the presence of or relative freedom from back flows of potentially contaminated air. BS 3202 has now been withdrawn, and a recently issued Draft for Development[2] proposes a quantitative standard method of assessing performance in terms of a newly defined concept of fume cupboard 'containment'. This is dependent on several parameters including velocity and is measurable by a proposed standard test procedure. Manufacturers now may, therefore, offer fume cupboards tested in these standard conditions and specify their potential containment capacities when operated at appropriate air flow velocities.

The new Draft for Development recognises the important influences of fume cupboard siting and layout, make-up air supply and additional room ventilation, as well as the need for adequate and reliable installations for the extraction ventilation of the fume cupboards. It offers recommendations in all these respects. Moreover, it recognises that users' requirements in respect of safety cannot be satisfied unless all these influences are taken into account and controlled in ventilated fume cupboard installations. Recommendations are given in *Part 2* for the minimum information to be exchanged between parties to contracts for the supply and/or installation of general-purpose fume cupboards as specified in *Part 1*. This valuable advice is logical and necessary if failures and indeed disputes are to be avoided, and the object of this chapter is to summarise the recommendations of *Part 2* and explain their purpose in the context of the responsibilities of manufacturers and users under Sections 6 and 2 respectively of the *Health and Safety at Work, etc. Act 1974.* This Act applies to laboratories as well as to workplaces of other kinds.

Figure 17.1 is a flow diagram for the information to be exchanged and shows the underlying logic of the procedure described in *Part 2* and which is perhaps not immediately evident on first reference. The first requirement is that the user should select a fume cupboard and an air flow rate considered safe by

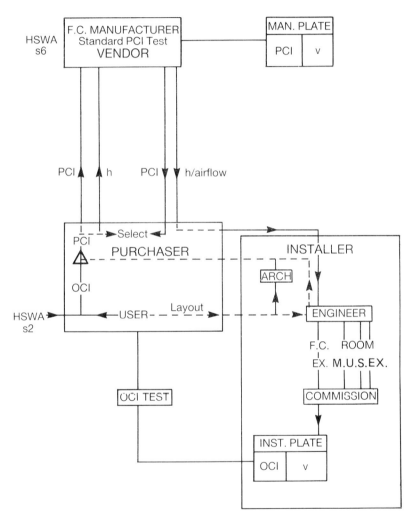

Fig. 17.1 – Flow sheet of requirements[2] for co-operative responsibilities for provision of a safe fume cupboard facility. Key: FC: fume cupboard; FC EX: fume cupboard extract; MU: make-up air; h: height of sash opeining; v: face air velocity; S: supply; EX: extract; HSWA: Health & Safety at Work Act; ARCH: architect.

the user in relation to the nature and amount of hazardous material to be released inside the cupboard and the levels of concentration likely to ensue outside the cupboard if the air is not totally contained. Three contracting parties are identified, two with direct responsibilities to the Health and Safety Executive (HSE) as manufacturer and user respectively, and a third, the installer/ventilating engineer, whose responsibilities lie in producing air flow conditions into the cupboard and in the room itself which will allow the user to use a potentially safe appliance in a safe manner.

The 'fume cupboard containment index' as defined in *Part 1* of the Draft is the ratio of concentration of contaminant inside the cupboard to the concentration at the sash opening. Other definitions included are 'potential containment index' (PCI) which is determined under standard idealised conditions in a test room free from air flows and obstacles that might otherwise disturb the flow into the working aperture, and the 'operational containment index' (OCI) determined in a real laboratory under typical working conditions.

Ultimately the user and the HSE are concerned with the concentration of possible contaminants outside the cupboard, which they will equate with Threshold Limit Values where appropriate. A procedure for calculating the necessary OCI and PCI values is recommended in *Part 3* (section 3.2) of the Draft. This incorporates a tentative and arbitrary factor to deal with air disturbances in the room, for example those caused by the air make-up system or which result from fume cupboard operation. Further research and experience is needed to assist in the choice of a factor appropriate to specific cases, but it is necessary to realise that this factor will rarely, if ever, be unity.

The factor is dependent also on design decisions made by several separate parties, who must therefore consult each other to obtain the best practicable combination of room layout, room and cupboard ventilation and fume cupboard design and air flow. They must communicate and record the details of the several particulars contributing to the overall performance standard of the installed fume cupboard. Figure 17.1 indicates the necessary exchanges of information recommended in the Draft.

The commissioning of the ventilating engineering systems will thus lead to the site testing of air flow performance in terms of face velocity into the sash. This will be recorded on an installation plate fixed to the cupboard. Similarly, where adequate containment cannot be interred from achieved velocity in relation to the manufacturer's test of the PCI, a site test of operational containment may be required to be carried out by the user or his agent. This also is recorded on the installation plate. The fume cupboard as supplied and type-tested will also bear a plate showing the PCI and velocity test performance.

The Draft for Development [2] points the way to the assessment and control of risk in the use of fume cupboards on a logical basis related to known data of the acceptable working levels of potentially toxic fumes and gases. It is the only method so far available which will allow estimates of protection to be predicted in the same terms as those used by industrial hygienists when assessing an actual working environment by personal monitoring. The Draft is to undergo a two-year trial period during which it is hoped it will be extensively used and assessed, and improved where necessary. Before such time as manufacturers are ready to offer data on the potential containment performance of these type-tested cupboards, users should not delay in carrying out containment tests on site. These will enable them to become familiar with the method and to explore the relationship of containment assessments with the numerical results of personal monitoring and the qualitative information provided by smoke testing.

REFERENCES

[1] BS 3202: 1959, *Recommendations on Laboratory Furniture and Fittings.* British Standards Institution, London, 1959.
[2] DD 80: 1982, *Laboratory Fume Cupboards,* British Standards Institution, London. 1982.

18

Reliability principles applied to ventilated enclosure systems safety engineering

K. C. Hignett and **B. A. Sayers**
Systems Reliability Service, United Kingdom Atomic Energy Authority, Culcheth

THE SAFETY AND RELIABILITY DIRECTORATE

Over the past twenty years, the Safety and Reliability Directorate of the United Kingdom Atomic Energy Authority (UKAEA) has developed methods of reliability analysis, supported by a comprehensive data bank. This reliability technology has been applied to nuclear installations with much success, as indicated by the very high safety record of the British nuclear industry. The National Centre of Systems Reliability now applies the technology to a wide range of non-nuclear engineering industries both in the United Kingdom and on an international scale through the Associate Membership of the Systems Reliability Service. This chapter introduces designers and users of laboratory systems to the principles of reliability assessments with particular reference to ventilated enclosures.

RELIABILITY

In discussing reliability it is important to have an agreed definition of a concept that is often expressed in purely qualitative terms. When comparing a number of pieces of equipment, or a number of systems, it is desirable to be able to quantify reliability, just as other parameters of performance, for example accuracy and repeatability, are quantified.

Thinking of these problems, if reliability can be considered on a comparative basis — for example, is this equipment more or less reliable than some other piece of equipment for a specific function, such as accuracy of operation — then the task becomes more simple and a definite conclusion may be drawn.

Further to this, it can be seen that reliability is invariably used as a parameter applied to a function or operation which has to be performed by a piece of equipment, a system or even a single component, and as such may be said to be a measure of the performance to do a specific task. Thus a measure of the reliability of a piece of equipment, for example a ventilated enclosure, may be the quality of the performance at the relevant time, or within a specific time interval. Timing must be brought into the concept since it is useless to carry out a function outside the time period during which it is required. Thus a general definition of reliability may be 'the chance or probability that a system performs in the desired manner within or for the period of time of interest'. A system can mean anything from a complete chemical plant to a component, or even, a human operator.

Reliability then is the opposite of failure, in that if a system does not perform in the desired manner, within, or for, the time period of interest, it may be deemed unreliable, or failed within its design intent.

In the case of a safety system, which may be one designed to protect personnel, the environment or the plant itself, failures may be either dangerous or safe. Dangerous failures jeopardise safety and generally leave the plant in an unprotected state. Safe failures usually generate spurious shut-downs, and so reduce the availability of the plant the safety system is designed to protect.

These two concepts, safety and availability, can now be considered in more detail.

SAFETY CONCEPTS

When considering safety, it is usual to think of the events, fire, explosion, contamination, etc., that can occur within a process and may lead to a hazard. To protect against such hazards, safety systems are installed which sense process faults and apply remedial action, usually in the form of shut-down, reduced production or alarms. A safety system thus awaits a demand for protective action from a process fault, which could lead to a potential hazard, and the user is then interested in the chances of the safety system working at this time, that is in its reliability or conversely, its probability of failure. The most critical forms of safety system failure are those which are not revealed during normal operation. An unrevealed failure is one which is not noticeable by the operators and is only brought to light by thorough and periodic proof testing. A failed safety system would exhibit normal working characteristics, but would not respond to a demand arising from a process failure.

The concept of safety can then be defined in terms of success or failure and is illustrated in Fig. 18.1.

AVAILABILITY CONCEPTS

Faults in safety systems which cause a shut-down of a plant unnecessarily, that is when a process demand is not present, will reduce the plant availability.

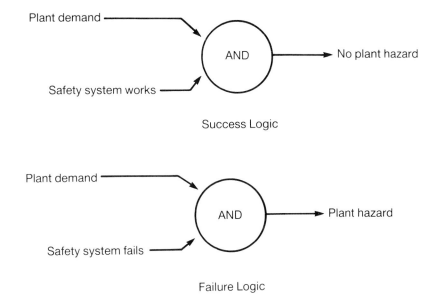

Fig. 18.1 — Success or failure of the concept of safety.

Generally such faults are revealed, in that the plant shuts down and in this case it is the number of times per year that the safety system fails in such a manner that is of interest and also how long it takes to effect a repair since the shut-down will continue in this state until the safety system is working again. The time to restart the plant will also be of importance. In the case where a safety system has shut-down a plant without a genuine demand, the plant could not be restarted, before safety system repair, without risking an unprotected demand. In some systems, availability studies are related to economics rather than safety, but in some other systems, the reverse is true as, for instance, in aircraft automatic landing systems.

Redundant safety systems, such as two safety systems instead of one, can be used to improve the reliability towards safety, in that if one safety system has failed dangerously (i.e. unrevealed) there is still the remaining safety system upon which to rely. However, having two safety systems instead of one will double the unwarranted shut-down rate, since either system can cause shut-down. This results in an economic penalty. This problem can be overcome by using 'majority voting' systems, where a process is shut-down not on the evidence of a single safety system but rather on a set number from the maximum, for example, two out of three, two out of four, etc.

Such voting systems give higher degrees of safety with lower spurious shut-down rates than simple redundant systems. They do, however, incur a penalty in increased maintenance costs due to the extra equipment required.

CAPABILITY AND RELIABILITY RELATIONSHIPS

Capability is the fundamental ability of a component, equipment or system to perform in the desired manner. Design capability may be inadequate due to designers' or manufacturers' errors.

Generally in reliability assessment studies, capability is assumed and any degradation of performance is attributed to unreliabilities in equipment, system, or human operators.

ASSESSMENT DEFINITIONS

Hazards

The hazard represents the incident or end event against which the safety systems are designed to provide protection. The occurrence of a hazard is dependent upon two events, which must be present at the same time:

(a) The plant must be out of control and proceeding towards a dangerous condition. This is termed the 'demand', because the condition demands protective action.

(b) The safety system must be in an unrevealed failed dangerous state when the plant demand occurs.

$$\begin{array}{ccc} \text{Hazard} \\ \text{rate} \end{array} = \begin{array}{c} \text{Process demand} \\ \text{rate} \end{array} \times \begin{array}{c} \text{Probability of failure of the} \\ \text{safety system on demand} \end{array}$$

Process demand

This is defined as an unwanted plant condition which indicates that the process is proceeding towards the defined hazard state. The annual rate of process demand should be the subject of a detailed assessment. Failures of equipment and human errors should be considered in addition to process conditions as contributing to the demand rate, referred to under 'Hazards'.

Failure probability

A safety system will fail to respond to a demand for protective action if a dangerous unrevealed failure is present in the safety system. Such unrevealed faults can only be revealed by means of proof testing which should be carried out at some regular, specified interval. The failure probability of a safety system is dependent therefore on the test interval and the failure rate of the system, and it can be shown that for a single system

$$P = 1 - e^{-\theta \tau}$$

where

P = cumulative probability of failure at time

θ = failure rate (exponential failure distrubution)

τ = proof test interval

when

$\theta \tau = \ll 1, P \approx \theta \tau$

BASIC MATHEMATICAL MODELLING

The mathematical model in this case is simply a logic diagram of the form shown in Fig. 18.1. In its total form the origins of the demands will be shown along with the associated safety systems failures. Each of the branches of the logic diagram will show the failure probability or applicable demand rate. The model will show where demands arise and where failure probabilities are high; it therefore presents a logical analysis of the system and shows the areas where improvements may need to be made. The end event of the model will normally be a hazard rate and this may be compared with a predetermined acceptable value which may be called the hazard criterion.

At the beginning of an assessment a fault tree is usually produced with the help of plant designers and operators. The fault tree forms the basis of the mathematical model showing the demand origins and whether protection is available for each demand. Fault trees are drawn to show how an undesirable end event is brought about and are usually drawn in failure logic, that is the failures are stated which lead to the undesirable end event. An example of a fault tree is shown in Fig. 18.2.

The complete mathematical model will show the demand rates or failure probabilities and will generate the system hazard rate on the basis of the logic depicted in the fault tree.

GENERAL CONCEPTS OF FUME CUPBOARDS

The main requirement for the effective operation of a fume cupboard is the provision of an adequate velocity of air entering through the working aperture. The problems encountered in the use of fume cupboards are of course, the differing requirements for differing aperture openings. For a simple extract system using a fixed damper position and a constant speed fan, the damper position would probably have been set to maintain the required air velocity with the maximum aperture opening. When the aperture is closed down, the velocity increases, which could lead to problems when dealing with fine particulates or powders. The ideal situation would be an automatically compensating system to maintain a constant velocity through any size of aperture. Numerous methods to achieve this ideal have been tried in the past, for example diaphragm valves, but they all tend to suffer from mechanical failures which could lead to hazardous situations. The application of vortex amplifier control appears to go a long way towards solving the problems of these modulating components, and takes the system towards the ideal of components with no moving parts, with the subsequent inherent high availability, due to very low failure rates. Additionally, the system then requires little or no maintenance.

The assessment of any laboratory ventilation system must consider the whole system in general and cannot just consider, say, the control of a fume

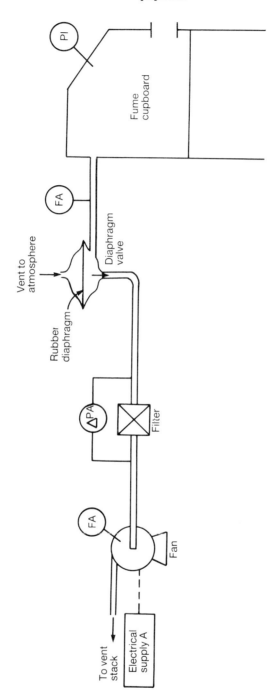

Fig. 18.2 — Conventional ventilated enclosure extract system. Key: FA — low flow alarm; ΔPA — high differential pressure alarm; PI — pressure indicator.

cupboard in isolation; the whole laboratory area must be taken as the whole system. This covers the immediate laboratory area adjacent to the fume cupboard, and also the area bounding the laboratory. The preferred air flow pattern should be from the outer laboratory area into the laboratory and then into the fume cupboard to be extracted by the fume cupboard extraction system. The laboratory ventilation system should not react adversely with the fume cupboard system.

The ventilation system can be divided into three:

(i)   Air into the laboratory and to the fume cupboard;
      Normally a small negative depression is required to limit the spread of fumes to other areas. However, clean condition rooms would require a small positive pressure to avoid the ingress of untreated air.

(ii) Air through the fume cupboard.
      This is not 100 per cent containment, but the system must maintain the minimum velocity requirements for the minimum to maximum position of sash openings. The entry of a fume cupboard should be designed on good aerodynamic principles to restrict any reverse flow patterns.

(iii) Air into the extract duct, and discharged to atmosphere.
      Extract ductwork should be as short as possible to ensure minimum pressure losses. It should incorporate suitable wash-down lines and drainage points, and should not allow any stagnant pockets to develop. The ductwork should be self-draining to a suitable treatment area. Ductwork should be suitably flanged to allow inspection and replacement. Location of the extract fans should be as close to the discharge point as possible to ensure that the majority of the ductwork is at negative pressures, and that leakage in the duct will result in a leakage into the system and not out of it. The discharge points should be sited to avoid any building interactive effects.

SPECIMEN COMPARATIVE HAZARD ASSESSMENT

Figure 18.2 represents a hypothetical but typical ventilated enclosure, which is held at sub-atmospheric pressure by means of a conventional diaphragm valve. The diaphragm valve acts as a pressure regulator by virtue of the displacement of the rubber diaphragm caused by the difference between atmospheric and fume cupboard pressures. Thus as the sash door of a fume cupboard is raised, then the increased fume cupboard pressure lifts the diaphragm, thereby increasing the flow through the valve and reducing the fume cupboard pressure towards its original depression.

The hazard that the designer is attempting to protect is the loss of depression in the fume cupboard pressure. A conventional gate valve would throttle the fan depression to a working level, but would only create a specified depression for a specific position of the sash door – as the sash was opened, the depression would drop. Using a diaphragm valve protects against the change in pressure in the fume cupboard caused by a change of position of the sash door. To assess the hazard rate of the system it is necessary to assess the failure rate of the safety system. A failure logic diagram for this system is shown in Fig. 18.3.

To find the total failure rate of the system in this case, where all but one failure leads to an OR gate, requires the addition of all the individual failure rates. The one AND gate in the diagram has an input of a probability and a failure rate and in this case the output is a straight multiplication of the two inputs to give a resultant failure rate. The final resultant rate of loss of depression in the fume cupboard is 0.6 times per year.

When this assessment has been done, a comparative assessment on a slightly differing system can be carried out to see if the overall failure rate can be improved. Substituting a vortex amplifier for the diaphragm valve, the system would look as in Fig. 18.4 whilst the failure logic diagram becomes as shown in Fig. 18.5. Then the resultant rate of loss of depression in the fume cupboard is 0.33 times per year. As can be seen, the use of a vortex amplifier in place of a diaphragm valve has achieved a reduction in the failure rate of the system, which the assessment has highlighted. Also from both of the assessments it can be seen which critical areas contribute the greatest amount to the failure rate of the system.

Thus, as can be seen, a major contributor to the overall failure rate of the system, shown in Fig. 18.4, is the fan failure rate. The provision of a second fan would alter the cost of the system significantly. However, it can be seen that the loss of electricity supply to the unit causes as many failures as failure of the fan itself. This suggests that the failure rate of the system can be improved by replacing the single electrical supply to the extract fan by duplicated independent electrical supply circuits, as in Fig. 18.6, at a possible lower cost. The resultant failure logic diagram is as shown in Fig. 18.7, with a system failure rate of 0.23 times per year.

CONCLUSIONS

Ways in which reliability analysis can be used to highlight critical areas of a system, and to indicate how improvements can be achieved by changing different areas have been described. improvements in availability (the proportion of time the system is fully serviceable) can be made in many ways, but changes can usually be classified under one of three headings:

(a) Reducing failure rates.
(b) Reducing down-time, that is the length of time an item remains out of service once it has failed.
(c) Introducing redundancy into the configuration of the system.

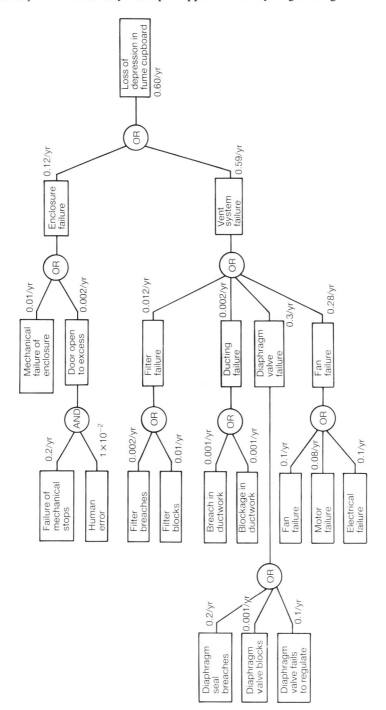

Fig. 18.3 — Failure logic diagram for a conventional ventilated enclosure extract system.

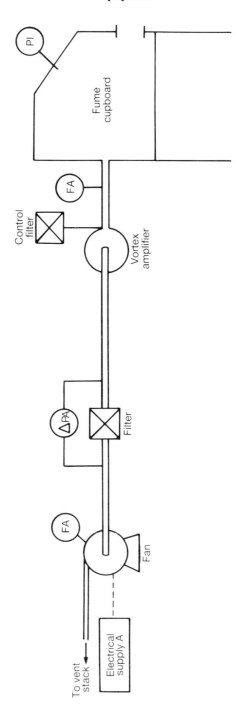

Fig. 18.4 – Ventilated enclosure extract system using a vortex amplifier. Key: FA – low flow alarm; ΔPA – high differential pressure alarm; PI – pressure indicator.

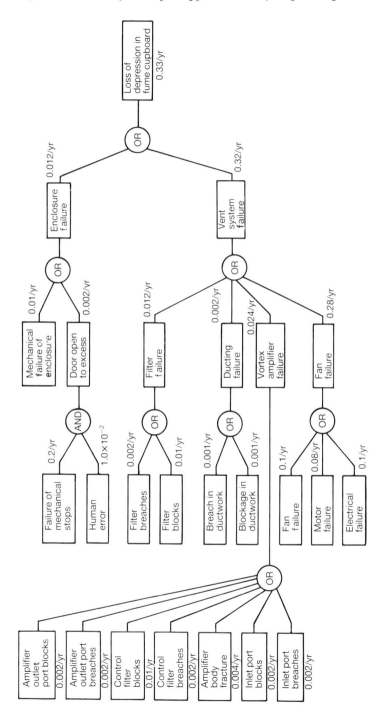

Fig. 18 5 — Failure logic diagram for a ventilated enclosure extract system using a vortex amplifier.

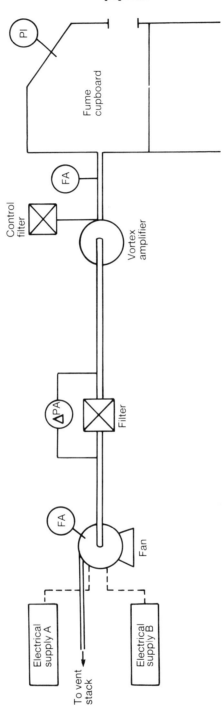

Fig. 18.6 – Ventilated enclosure extract system using a vortex amplifier and two electrical supplies in redundancy for the extract fan.
Key: FA – low flow alarm; ΔPA – high differential pressure alarm; PI – pressure indicator.

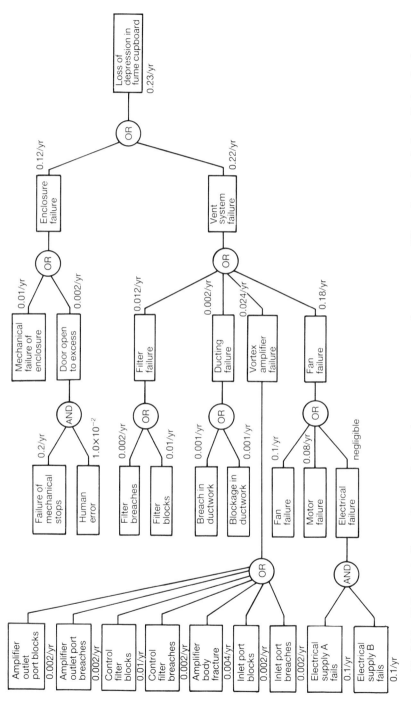

Fig. 18.7 – Failure logic diagram for a ventilated enclosure extract system using a vortex amplifier and two electrical supplies in redundancy for the extract fan.

Changing the diaphragm valve for a vortex amplifier is an example of changing one item of equipment for another with a reduced failure rate. Providing duplicated independent electrical supply circuits to the fan units is an example of introducing redundancy into the system configuration.

Duplicating the fans would be the next major step in improving the reliability of the system still further. In the system described this may be considered an unnecessary expense, but in a more extensive system, where several fume cupboards were served by a common extract system, duplication of the fans would be a natural and important step forward. If automatic changeover facilities were provided, this would then be an example of standby redundancy.

Many other possible design improvments to this very simple system could obviously be suggested, but the advantage of a quantified reliability assessment is that it is possible to identify the most important ones, concentrating design effort where it is seen to be most worthwhile.

FURTHER READING

A. J. Bourne, *The National Centre of Systems Reliability and some aspects of its Future Activities,* NCSR Report RI, UKAEA, 1975.

A. E. Greene and A. J. Bourne, *Reliability Considerations for Automatic Protective Systems,* Report AHSB (S) R91, UKAEA, 1965.

A. E. Green and A. J. Bourne, *Reliability Technology,* Wiley—Interscience Publishers, 1972.

G. Hensley, *Plant and Process Reliability,* NCSR Report SRS/CR/1, National Centre of Systems Reliability, UKAEA, 1976.

The Systems Reliability Service Data Bank, United Kingdom Atomic Energy Authority, Risley, Warrington.

R. M. Stewart and G. Hensley, *High Integrity Protective Systems on Hazardous Chemical Plants,* Paper presented to the European Nuclear Energy Agency Committee on Reactor Safety Technology, Munich, 26—28 May 1971.

The Flixborough Disaster – Report of the Court of Inquiry, HMSO, London, 1975.

Health and Safety at Work etc. Act 1974, Public General Acts, Eliz. 2, ch. 37, HMSO, London, 1975.

19

The design of laboratory furniture for government establishments

P. Boardman

Design and Furnishing Group, The Crown Suppliers

THE DESIGN AND FURNISHING GROUP

The Group is part of the Property Services Agency (PSA) which is responsible to the Department of the Environment. Although part of a government department, PSA Supplies, known since January 1984 as 'The Crown Suppliers', operates effectively as a business organisation and is engaged in the supply of a wide variety of goods and services essential to the working of the government estate. It is an independent unit of accountable management, financially self-supported from the sale of goods and services. Although operating on commercial lines its purpose is not necessarily to make a profit, but to ensure that customers who spend public funds receive true value for money.

An important but relatively small part of the service provided by the Design and Furnishing Group is related to laboratory design and servicing. Laboratory furniture and equipment is supplied to various government departments involved in a wide variety of work, these include the Ministry of Agriculture, Fisheries and Food, the Ministry of Defence and the Laboratory of the Government Chemist (LGC). Other bodies in the public sector, such as hospitals, universities and local authorities, are also supplied through central contracts which are at special terms on a call-off basis. Some of these organisations occupy laboratories which are not totally suited to their activities because of limited building facilities; others are in purpose-built establishments. The Group must therefore cater for a wide variety of situations.

The advantages of this arrangement to both The Crown Suppliers and its clients are that

(a) the design service provided is based on the practical experience of a field force;
(b) a wide range of goods and services of consistent quality is available to clients; and

(c) because of the large quantities involved, advantage can be taken of com-
petitively arranged bulk purchase contracts with consequent good value for
money.

DESIGN AND TRENDS IN LABORATORY FURNITURE

Until the early 1970s the furnishing of each laboratory was treated as a one-off
special job in which furniture built in as a fixture to the building was used, solid
teak benches 1.5 m in width were provided, fixed to purpose-made under-bench
units and service casings. These were constructed from top quality materials
appreciated by joiners and cabinet makers and made to last in the best tradition
of furniture manufacture. These traditional methods often created problems.
The scientists, chemists and engineers had a laboratory of a design which made
later changes, to accommodate new activities, often difficult and expensive.

During 1974/75 PSA conceived the idea that this problem could be alleviated
in an economical and practical way by the use of a standard range of free-
standing modular furniture. This concept offered the advantage of the use of
mass production techniques with good quality control and provided PSA field
force staff with a range of standard components which clients could evaluate,
select and plan within their available laboratory space. Having selected the
range of furniture suitable for their requirements and prepared the site, installation of modular components would take less time to install and create less
disruption to adjacent laboratories because of the reduced period required for
site work.

In addition to this range of modular furniture, a basic range of service casings
was designed to provide the local field force with common data and components.
Site contractors soon became aware of PSA requirements and a working pattern
was gradually established. Most clients have now accepted this basic concept as
the norm, and have reported positive advantages particularly when access is
required for additional services or, in emergencies, when there is a need to move
furniture to clear up spillages.

To meet the needs of laboratories which have special requirements the
expertise of The Crown Supplier's mechanical and engineering staff and the PSA
Department of Mechanical and Electrical Services (DMEES) laboratory is available. This enables a full service to be provided that extends far beyond that
which normally comes under the heading of furniture supplies. Thus The Crown
Supplier's engineers have developed and arranged a contract for the supply of
safety cabinets conforming to the three classes specified in the relevant British
Standard Specification[1], and have recently produced a purpose-made solvent
storage refrigerator which is expected to be adopted by the trade as a standard
design.

Contracts have now been placed which provide a standard range of laboratory
furniture suited to most situations and consisting of some 134 items. These

include a variety of benches, pedestal storage units, sinks, service casings, protective screens, fume cupboards, workshop benches, and fire resistant cabinets. The laboratory benches are free standing, supported by a steel frame and are available in two standard heights, two depths and three widths to make up any length required. Tops are produced in three finishes — teak, plastics laminate, with a choice of three colours, and linoleum. Flexible PVC skirting allows the benches to be pushed tight to service casings or wall to form a water-resistant seal. Under-bench pedestals are either mobile or suspended from the bench frame and are produced in two widths. Service casings have pre-drilled holes in a number of permutations to accommodate all service outlets. Benches are delivered in knock-down form, but all pedestals and storage units are fully assembled.

Service trunking has also been developed specifically for instrument benches to provide the necessary electric and industrial gas supplies. Another specifically designed item that inter-relates to other items within the range is a wooden under-bench introduced for use in British Telecom laboratories. More recent additions to the range are the factory-made, fixed and mobile service casings. These mobile service casings set the trend for future developments, to provide full mobility and/or interchangeability in re-arranging the laboratory to suit differing applications with minimum loss of laboratory working time. It is also envisaged that in future developments the mobile service casing will be provided complete with electrical, water and gas services, etc., with a simple jointing system so that benching and services can be re-arranged at short notice, by local maintenance staff or laboratory technicians.

FUME CUPBOARDS

Some years ago LGC, which uses over 200 fume cupboards for a wide range of varying applications, was unable to find, from commercial sources, a fume cupboard that could be considered suitable for universal use. The main requirements were that it should

(a) be a simple and impervious one-piece chamber which was smooth, without join lines, and easy to clean;
(b) have good chemical and acid resistance;
(c) have good visibility inside the unit;
(d) have a sink base that could be covered with a removable work surface;
(e) have reliable service fittings for gas, water, compressed air and electricity;
(f) have a simple and effective means of adjusting sash height; and
(g) be a unit that could be simply and easily installed and maintained.

Traditionally, The Crown Suppliers had provided purpose-built wooden fume cupboard with a glass sash. LGC set the Group a challenge of designing and producing a unit to meet this brief.

The first cupboard conceived was made from glass-reinforced plastics (GRP) and included some wooden components. Stage by stage a complete one-piece GRP moulding was evolved which is considered to be second to none in design. This fume cupboard which has been in service since 1980 has an excellent interior finish in fire- and chemically-resistant GRP. The excellence of finish and quality of materials at this stage is not questioned and has proved very successful under extreme usage for work involving digestion with fuming acids. One such cupboard where forty 100 ml beakers of mineral acid are fuming continuously on an electric hotplate shows no signs of deterioration. This success, however, depends on good housekeeping and a good extract design; staining and contamination will occur if undiluted condensates are allowed to drain back into the fume cupboard. In these conditions it is recommended that a condensate collector with a wash-down nozzle should be incorporated. Correctly fitted this will prevent both condensates and water from the wash-down facility from draining back into the working area of the fume cupboard.

In anticipation of the recommendations put forward in the British Standards Institution Draft for Development[2] most fume cupboard manufacturers have considered the provision of various air-flow failure warning devices and aerofoil sections to improve both air flow into the fume cupboard; and containment. They are also considering methods to combat problems related to high face velocities when the sash of the cupboard is at its lowest working position, to improve energy conservation, to dilute fumes discharged into the atmosphere and to protect the extract system from the effects of corrosion. The Crown Suppliers, in common with most designers and manufacturers, is well aware of these problems and a number of solutions is being considered.

A cooperative programme between The Crown Suppliers, DMEES, LGC, the Health and Safety Executive and other bodies experienced in this field, has gathered information to produce an effective compromise. This work has shown that to correct the problems faced in relation to fume cupboards it is necessary to resolve the problems connected with the extract system. It is now possible to provide, in some cases, the cupboard and extract system as a complete package.

New projects can be divided into basic types, those where the client is carrying out special or secret work and is well aware of what is required and prepared to provide specific details, and those where the client is able to use one of the standard PSA range of cupboards.

In the first case a performance specification is prepared and offered to the trade for competitive tender. These are evaluated by PSA and supply and delivery arranged. In the second case, PSA is some way towards being able to offer the client a system which is either:

(i)   a basic fume cupboard of well proven construction and materials; or
(ii) the same fume cupboard incorporating modifications to the internal shape
      of the chamber, with a back baffle and external aerofoil sections to the cill

and sash which ensures an even flow of air through the cupboard and eliminates the vortices normally apparent in smoke tests.

The standard range of fume cupboards now available is in 1200 and 1500 mm widths in single- and double-sash versions. The latter is available where the laboratory ceiling height is insufficient to accommodate a single sash. Although the fume cupboards are generally manufactured in GRP, a stainless steel inner chamber is available when it is considered that solvent usage or high temperatures would result in deterioration. A new range of fume cupboards will include the facility of a sash stop and a face velocity indicator which will give clear indication of a reverse flow situation when insufficient make-up air is available or where additional apparatus may take air from the laboratory.

The improved design fume cupboard can be assembled on site in approximately half an hour and also provides, for maintenance purposes, access to service valves, etc., without disturbance to adjacent units. Top quality *Mark Lab* control valves are purposely positioned below the cill front to avoid disturbing the flow of air into the cupboard. The control valves are colour coded and shaped to DIN standards.

It is hoped that future developments will widen the scope of the fume cupboard supply contract to include a basic PVC trunking system coated with GRP, as recommended by LGC, to provide structural rigidity and fire resistance. The system will also include pre-moulded flanges, bends, T-pieces and flexible joints that will provide a complete package free from any of the solvent joints and welds that have, in the past, proved to be a weakness in maintaining reliable seals.

A major problem with the design of the conventional fume cupboard has been that of air flow. Unless sufficient make-up air is supplied within the laboratory, then cold draughts, banging doors, excessive noise, etc., can limit the flow of air to the fume cupboard. Three problems need to be resolved with regard to the extraction of air from a fume cupboard.

(1) The high cost of energy lost by extraction of warmed laboratory air through the cupboard.

(2) The speed of air movement at lower sash positions which can upset experiments inside the cupboard.

(3) The maintenance of dilution of fumes discharged to the atmosphere (i.e. the efflux factor). An exhaust velocity of 10 to 13 m/s from a stack at least 3 m in height is normally recommended. This is essential for the health and safety of people and property in areas surrounding the laboratory.

One solution to the second problem is the use of a fume cupboard with a by-pass system. These units are very expensive in energy, only partly efficient and present a fire or blow-back hazard in certain situations. The Crown Suppliers have moved sufficiently towards development of this type of fume cupboard to be able to include it in its standard range, should a need for it arise in the future.

Development work to overcome two of the problems includes:

(a) Electronic controls to vary fan speed and damper positions. These require switch gear and integrated circuitry which, because of their sophisticated nature put into question reliability and consequently safety. Installation and maintenance can also be costly.

(b) A vortex control being developed under contract for PSA by the United Kingdom Atomic Energy Authority. The feasibility study, indicated that vortex control could act as a linear damper so that when the sash was raised or lowered a constant face velocity would automatically be achieved without any additional mechanical or electrical devices being introduced into the system. However, current work has shown that although the concept is possible there is insufficient signal to the control port of the vortex controller and, therefore, an additional air flow is required to make it operational. This makes the idea less attractive but nevertheless a solution along these lines is possible.

Customer feed-back to The Crown Suppliers, as to all suppliers, will help provide better installation in the future. All answers may not be available today but a central purchasing organisation can provide a focal point to evaluate and rationalise with a view to future provision of a high quality service and goods at extremely competitive prices.

When planning to build or refurbish a laboratory, one should think first of what it could be used for in ten years time or in the event of emergencies. It should be planned with flexibility in mind, and ideas discussed with the engineers and architects at an early stage because they can only resolve problems that they are made aware of at the planning stage. It should be ensured that they know about the built-in adapatability of flexible laboratory furniture and fume cupboards as these will help them to provide a safe and efficient laboratory environment at an economical cost.

REFERENCES

[1] BS 5726:1979, *Specification for Microbiological Safety Cabinets,* British Standards Institution, London, 1982.

[2] DD 80:1982, *Laboratory Fume Cupboards,* British Standards Institution, London, 1982.

20

New innovations in materials used in the manufacture of laboratory furniture

L. G. Haley
Cygnet Joinery Ltd

LABORATORY BENCH TOPS

Traditional laboratory benches were constructed with solid Burma teak (*Tectona grandis*) tops, supported by either polished hardwood under-bench storage units or hardwood table frames. In wet areas and for tops which had to contend with spillage, solid timber or laminated timber tops were covered with chemical sheet lead. This material was also used as a lining for sinks and drainage channels. The demand for good quality Burma teak greatly exceeded its availability, and consequently its cost increased far more rapidly than most other hardwoods. Square edge boards suitable for the manufacture of bench tops also became increasingly difficult to obtain. Laboratory furniture manufacturers were faced with the problem of obtaining suitable alternatives. The hardwoods iroko, afzelia and afromosia proved to be excellent substitutes for teak providing they were kiln-dried to the appropriate kilning schedules. These hardwoods cost around £300/m^3 compared with approximately £1000/m^3 for teak. Clear lacquer finishes with epoxy or other synthetic resin bases increase the chemical resistance of these timbers. When choosing a finish for hardwood tops it is advisable to take into consideration the natural movement which takes place in solid timber tops and use a lacquer which can cope with this movement without developing surface cracks.

Melamine laminates are now used extensively in the manufacture of laboratory bench tops. The laminate sheets, usually 1.3 or 1.5 mm in thickness, are bonded to chipboard or plywood core material. If tops are perforated for sinks or drip cups, moisture resistant chipboard or exterior grade plywood should be used. Stainless steel sink bowls with a flat flanged rim can be bonded with an epoxy resin adhesive directly to the underside of the laminate sheet which is then trimmed to the internal profile of the sink bowl. Most laminate sheets are now produced with non-reflective finishes, and they are also available in fire retardant finishes. Standard sheet sizes are 3050 × 1220 mm or 4120 × 1530 mm. Work tops with post-formed front edges eliminate the need for applied lippings. The sheet can be formed to a 12 mm external or internal radius, and post-formed rear upstands can be incorporated.

Solid grade laminates are being used increasingly for laboratory bench tops. This material is available in a wide range of surface colours with a standard brown core and thicknesses varying between 3 mm and 25 mm. There is a standard interior grade, an exterior grade, a flame retardant grade, and an acid resistant grade which is only available in black with a black core. Dished work surfaces can be formed in solid grade laminate tops by bonding 12 mm thick margin pieces, with chamfered inner edges, to 16 mm or 18 mm flat sheets. Where long tops are required, solid grade laminate sheets, 3050 × 1530 mm, can be jointed with loose tongue joints in a similar manner to hardwood bench tops.

Chipboard or plywood tops can be covered with sheet polyvinylchloride or polypropylene to provide a wet area work surface resistant to a wide range of corrosive chemicals. Polypropylene sinks, drip cups and drainage troughs are produced as standard items and are easily installed in most types of work tops.

The use of stainless steel in laboratory furniture has increased considerably in recent years with the main applications being in work tops, sinks and sink bowls. The most corrosion resistant quality in general use is specified as EN 58J, FMB or 316 quality. As a safety precaution earthing tags should be fitted to all stainless steel tops and sinks.

Cast epoxy resin bench tops are now manufactured in thicknesses of 19, 25 or 32 mm and up to 2438 mm in length and 1372 mm in width. They are available as plain flat tops or complete with an integral rear upstand; corner sections are available. Sections of top can be jointed using an epoxy resin cement. Cast epoxy sinks, drainage troughs and drip cups are produced as standard items together with dished work surfaces suitable for use in fume cupboards. Epoxy resin bench tops are highly resistant to corrosion from exposure to common laboratory reagents and solvents. The hard non-porous surface is also stain resistant and the standard colour is black. When using epoxy resin bench tops it is advisable to ensure that all machine operations required have been carried out before tops are despatched from the factory. Recent tests have shown that it can take 35 minutes using a tungsten carbide tipped drill, to make a 25 mm hole through a 32 mm top!

Heat resistant bench tops are produced from a mixture of calcium silicate binders reinforced with mineral fibres. They stain very easily but various coatings can be applied to prevent this and also increase the chemical resistance. The standard sheet size is 2400 × 1200 mm and thickness varies from 6 to 32 mm. Epoxy jointing compound is recommended for sealing joints in large bench tops.

Ceramic tiles can be fabricated into laboratory work tops making them resistant to most acids, alkalis and organic solvents at temperatures up to 120°C. There is a wide choice of tiles available and a popular size is 150 × 150 × 13 mm.

The tiles should be bedded with 3 mm joints between and pointed with chemical resistant cement. To ensure strength and stability each top should be reinforced with steel and glass reinforced plastic polyester resin.

A new type of plastics worktop was introduced by Du Pont in America about 12 years ago. This material is a filled methyl methacrylate polymer and produced in cast sheets 6, 12 and 18 mm thick, available in three standard colours. Sink bowls in the same material are available as standard items and can be fabricated as an integral part of the work top. These tops have been installed in several laboratories at University College London where spillage of bio-hazardous materials was considered likely to cause problems. The material can be worked with power hand tools, providing tungsten carbide tipped cutters are used.

UNDER-BENCH STORAGE UNITS

The three most common types of under-bench storage units are:

(1) The fixed unit which directly supports the bench work top.
(2) The removable unit which is a free standing unit capable of being repositioned by laboratory staff to suit individual requirements.
(3) The suspended unit which provides similar facilities to the removable unit, but is suspended or supported usually between 200 and 300 mm above the floor level. This type of unit reduces the storage space available below bench top level but simplifies cleaning problems and mopping up operations when spillages occur.

In the case of removable and suspended units, independent bench top supports are required, normally in mild steel and fabricated in the form of cantilever frames or table frames. Bench stability can be achieved easily using tubular metal frames and rails without resorting to wall or floor fixings, and they can be fitted with adjustable feet for levelling purposes. Various finishes can be applied to the metalwork but two that have proved successful are epoxy powder coating and bright zinc plating.

The use of solid timbers in the manufacture of laboratory storage units has declined over the years, the current trend being to use metal or man-made board materials, together with plastics trays. However, it should be noted that the developments mentioned in this chapter refer to timber laboratory units. Eighteen mm chipboard is used extensively for carcases, doors and drawer fronts of laboratory storage units. It can be veneered with hardwood or plastic laminate sheets; exposed edges should be lipped. Doors and drawer fronts are often veneered with post-formed melamine laminate sheet formed to a radius on either the two vertical or the two horizontal edges, eliminating two of the four lippings required on square edge doors and drawer fronts which are susceptible to damage. Exterior grade plywood, 18 mm thick, is an excellent alternative and is superior to chipboard in many respects, but the cost of this material is approximately £8/m² against £1.65/m² for plain chipboard, which explains the reason why plywood is rarely used as a core material for laboratory units. Medium density fibreboard costs roughly 50 per cent less than plywood, and is a precision-made resin bonded wood fibreboard with a uniform density from surface to core. Dimensionally it is stable and its high internal bond strength gives substantially greater screw holding over chipboard. The surface is suitable for the application of veneers, and is also an excellent surface for paint or lacquer finishes. Edge lipping is unnecessary, therefore the edges of doors or drawer fronts either square or with radius can be painted in a similar manner to the surface. Coloured, acid-catalysed lacquer applied by the curtain coating process to medium density fibreboard produces an excellent finish at an economical cost and is easily repaired if damaged. The use of trays for storage purposes in lieu of cupboards and drawers is sometimes preferred. Timber, wire or plastics trays are used, usually housed in a tray unit which can be open or fitted with doors.

An American company produces a range of furniture units which consist of bench tops supported by metal complete with metal racks from which various depths of plastics trays are suspended. Using this system the whole of the under-bench storage consists of open plastics trays. The bench units are free standing and are also available as mobile units mounted on castors.

LABORATORY FUME CUPBOARDS

During the past few years a considerable amount of discussion has taken place on fume cupboard design.

The British Standards Institution recently published a Draft for Development of laboratory fume cupboards[1], prepared under the direction of the Laboratory Apparatus Standards Committee, which supersedes Section 3 of BS 3202:1959, *Recommendations on Laboratory Furniture and Fittings.* It covers, in three Parts: safety requirements and performance testing; recommendations for information to be exchanged between purchaser, vendor and installer, and recommendations for installation; and recommendations for selection, use and maintenance. The publication must not be regarded as a British Standard. It is issued in the Draft for Development series and is of a provisional nature, because it features the novel concept of a containment index and describes methods for its measurement.

REFERENCE

[1] DD 80:1982, *Laboratory Fume Cupboards,* British Standards Institution, London, 1982.

21

Water supplies for laboratories

N. H. E. Hanby
Southern Water Authority, Hampshire Division

INTRODUCTION

The content of this chapter is based on the water supply practices adopted in the United Kingdom as expressed in two publications from the Department of the Environment[1, 2]. Within the United Kingdom there are variations in practice but there are greater variations between the United Kingdom and countries overseas. It is therefore important that the local Water Authority is informed of the details of a proposed installation so that they may advise on any changes necessary to satisfy local requirements.

THE STORAGE CISTERN

Supply

It is important that the water supply to a laboratory be taken from a storage cistern on the premises for two reasons. The risk of toxic, infective or otherwise objectionable material from the laboratory being siphoned back into the public supply is avoided. Continuity of the supply to the laboratory is ensured so that experimental work of long duration is not upset by an interruption in the water supply caused by a burst main or other problems with the public supply system.

Capacity

As a general rule, the capacity of the storage cistern from which the supply is to be taken should be sufficient to provide water for one working day. It is usually desirable to err on the side of generosity when assessing this quantity as the demand for water tends to increase with the passage of time and it is usually difficult to increase water storage capacity, once the laboratory is working.

Byelaw requirements

There are a number of Byelaws that apply to the storage cistern of which the most important are:

(a) The entry of water into the cistern should be controlled by means of a float-operated valve which should comply with the requirements of the relevant British Standard Specification[3,4].

(b) The cistern should be equipped with a suitable overflow which should discharge in a conspicuous position.

(c) The cistern should be so placed that the interior can be readily inspected and cleansed.

(d) The cistern should be fitted with a properly made cover which should be rigid, should exclude light and should have turned-down edges to hold it in position.

(e) The cistern should be satisfactorily supported. For cisterns made of plastics material this probably means a continuous surface over the full size of the base.

Location

The location of the cistern in a building will be determined by the water pressure available from public mains and by the construction of the building. Three different situations are illustrated.

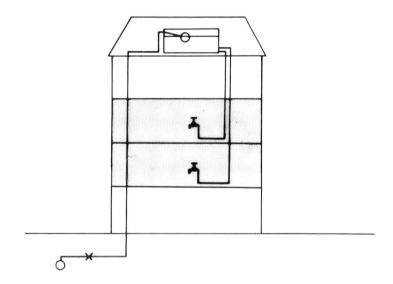

Fig. 21.1 – Roof cistern supplied directly from water main. Taps supplied by gravity.

(1) Sufficient water pressure and roof space (Fig. 21.1).

Provided that the structural strength of the building is adequate the most satisfactory arrangement is to place the storage cistern in the roof space and supply it directly from the public main. The water is supplied, from the storage cistern, to the separate floors of the building by means of distributing pipes. A separate distributing pipe should be installed for each floor to prevent siphoning back from the upper to lower floors.

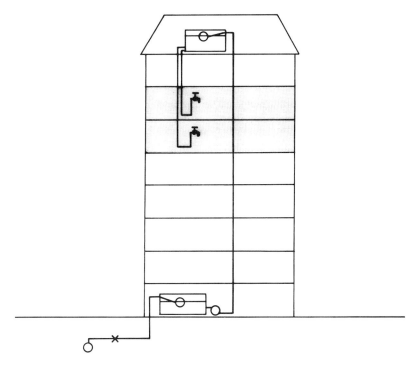

Fig. 21.2 — Roof cistern supplied by pump drawing from a low level cistern. Taps supplied by gravity.

(2) Insufficient water pressure but sufficient roof space (Fig. 21.2).

The supply is first taken into a storage cistern at low level, probably on the ground floor. Water is raised from this cistern to the roof cistern by means of a pump. To ensure reliability, two pumps should be provided and arranged so that one pump performs the duty for two weeks followed by the second pump for one week. If this arrangement is maintained, the share of duty performed by the first pump will be double that performed by the second which can therefore be expected still to be in good condition when the first is affected by wear and tear. Distribution to the separate floors of the building is as in (1).

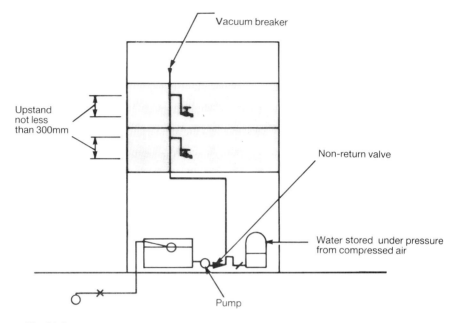

Vacuum breaker

Upstand
not less
than 300mm

Non-return valve

Water stored under pressure
from compressed air

Pump

Fig. 21.3 – Taps supplied from a pneumatic storage system fed by a pump drawing
from a low level cistern.

(3) Insufficient water pressure and insufficient roof space (Fig. 21.3).

The supply is first taken to a low-level cistern with sufficient capacity
to provide at least a full day's supply. Water is supplied to the separate
floors by means of a pump drawing from the cistern and delivering to the
various outlets. With this arrangement, economy in pump running time can
be achieved by providing a pneumatic storage vessel which should be connected to the pipework on the delivery side of the pump.

To guard against siphonage back from floor to floor and to the storage
cistern, a suitable non-return valve should be installed adjacent to the delivery
connection of the pump, and a vacuum breaker at the highest point of the
pipework. In addition, each run of pipework supplying outlets should be
connected to the pump delivery pipe at a height which is not less than
300 mm above the level of the draw-off points.

To ensure reliability, two pumps should be provided with performance
arrangements as in (2).

PIPEWORK

Materials

The most widely used material for pipework for water within buildings is copper,
with joints of either the soldered capillary type or the compression type. It is

advisable to make use of the latter type if the pipework is likely to be altered at any time. Although more expensive initially, compression joints can be dismantled and re-used. An alternative material is stainless steel which may be used in much the same way as copper, with soldered capillary, or compression joints. A special flux is required when making soldered joints on stainless steel tube.

Galvanised steel is likely to give rise to corrosion problems unless the water in the area is stable when in contact with galvanised steel surfaces. In hard water areas there is a tendency for zinc carbonate to be formed. This takes the form of small grains of a gritty texture which tend to lodge in float operated valves, taps, meters, etc., and prevent correct functioning. Plastics materials are becoming readily available and certain grades of plastics piping are suitable for use in hot water systems.

It is important to avoid mixing different metals in the same pipework installation. Dissimilar metals in contact can form galvanic cells which lead to consequent corrosion.

Identification of different water systems
It is desirable to distinguish between different water systems in situations where pipework installations are likely to be complex, by the selection of different pipe materials. Stainless steel could, for example, be used for a drinking water supply, copper for the hot water supply and plastics tube for cold water not required for drinking.

Selection of water pipes and fittings
The Water Research Centre (WRC) publishes a Directory [5], listing the various materials, fittings and appliances which have been examined by the WRC Testing Station and found to satisfy the requirements of the Water Byelaws.

WATER CONDITIONERS
In hard water areas the formation of lime scale in pipework, boilers, distillation apparatus, etc., can be a problem and in some cases, such problems can be alleviated by the use of a magnetic water conditioner. The mechanism by which such devices operate is not clearly understood, but when used in conjunction with certain types of hard water, they lead to the formation of a soft sludge instead of a hard scale. In the author's experience such an improvement has been found to occur when a magnetic water conditioner is used with water derived from chalk boreholes.

REFERENCES
[1] *Model Water Byelaws*, 1966 edition, HMSO, London, 1966.
[2] *Report of the Committee on Backsiphonage in Water Installations*, HMSO, London, 1974.

[3] BS 1212, Part 2: 1970, *Specification for Float Operated Valves (excluding floats), Diaphragm Type (brass body)*, British Standards Institution, London, 1980.

[4] BS 1212, Part 3: 1979, *Specification for Float Operated Valves (excluding floats), Diaphragm Type (plastics body)*, British Standards Institution, London, 1979.

[5] *Directory of Water Fittings and Materials*, United Kingdom Water Fitting Byelaw Scheme, Water Research Centre, Slough.

22

Lighting for laboratories

P. R. Boyce
Electricity Council Research Centre, Capenhurst

THE PROBLEM FACING THE DESIGNER

Every laboratory has its own characteristics and functions. The archetypal laboratory shown in Fig. 22.1 has fixed island beaches with chemicals and services, shelves lining the walls and the occasional piece of larger equipment. This is only one of a number of different types of laboratory and Fig. 22.2 shows an industrial laboratory in which the melting characteristics of arc furnaces are examined. Between these two extremes lie laboratories for physics, chemistry, biology, metallurgy, zoology, pathology, etc., each having its own characteristics. This diversity in what constitutes a laboratory ensures that there is no single solution to the problem of lighting a laboratory. The designer needs to study carefully the work to be carried out there and then consider what lighting is available to meet the requirements.

The requirements can be considered under three headings:

(1) Task performance.
(2) Avoiding discomfort.
(3) Operating conditions.

Task performance

Laboratories exist for scientific work to be done in safety, with speed and precision. Lighting is always necessary for safety reasons when people are in a laboratory, but lighting of the right quantity and quality can also facilitate the performance of laboratory work. The first task facing the lighting designer is to establish the visual requirements of the work to be done. This involves considering the size and contrast of the details that have to be seen, the extent to which movement has be be perceived, whether shape or form has to be clearly revealed and the precision with which colours have to be distinguished. Different tasks will have different combinations of these elements and each element is influenced

Fig. 22.1 – The archetypal laboratory.

by a different aspect of the lighting conditions. The illuminance on the task influences the visual acuity and contrast sensitivity of the visual system and hence the degree of detail that can be seen. The illuminance on the task and its uniformity over the task area affects the detection of movement. The distribution of light creates luminance patterns, highlight patterns and shadow patterns on objects and these determine how the objects appear. The spectral composition of the light emitted by the light source influences the ease with which colours can be discriminated. The lighting designer's job is to match the attributes of the lighting to the requirements of the task.

Avoiding discomfort
Although ensuring people's safety and facilitating task performance are the major functions of the lighting installation in a laboratory, it is also necessary to ensure that the lighting does not cause discomfort. Discomfort is caused whenever the visual system has to operate beyond or close to the limits of its capabilities. The discomfort will make itself known by the occurrence of one or more of the

Fig. 22.2 – An industrial laboratory for the study of the melting properties of
arc furnaces.

symptoms of eyestrain, which range from those which directly affect the eye,
for example sore, itchy eyes, to the more remote generalised symptoms such as
headaches. This in turn suggests that discomfort can arise from such diverse
sources as the muscles of the ocular motor system or from the brain itself when
it attempts to interpret small, blurred retinal images.

The lighting conditions which influence discomfort can be divided into
those that operate directly on the visual system and those that cause discomfort
indirectly because of the characteristics of the task to be performed. The three
aspects of lighting which cause discomfort directly are the luminance to which
the visual system is adapted, the range of luminances which occur simultaneously
in the interior, and the extent to which visible flicker occurs. To deal with the

last first, flicker is due to instability in either the light source, the control gear or the power supply. If flicker is visible to any marked extent it should be eliminated by changing lamps and/or control gear or by stabilising the power supply. The range of luminances which occur simultaneously in an interior is a function of the distribution of light from the lighting installation and the reflectances of the interior surfaces. Where the range of luminances is excessive the higher luminances will be considered glaring. It is also possible for the whole interior to be too bright in which case discomfort will again occur and can only be cured by reducing the overall luminance in some way.

Fig. 22.3 – Veiling reflections on a self-luminous digital scale, caused by the reflection of a luminaire in the glass cover.

Indirect discomfort is almost always a result of the interrelationship between the task characteristics and the lighting conditions not being considered carefully. Any task in which the detail which needs to be seen is not clearly visible will cause visual discomfort. The cause of the low visibility can be simply too low a task illuminance. It can also be veiling reflections from the task, when the task is a specular reflector. The most usual situation in which veiling reflections occur is when a high luminance source such as a luminaire or a window is reflected from a specular surface; the high luminance reflection masks the material which it is necessary to see (Fig. 22.3).

Operating conditions

In addition to considering the effects of the lighting on people in terms of task performance and discomfort, the lighting designer also has to assess the condditions in which the lighting will have to operate. These conditions may be described as hazardous or hostile. For example, in some laboratories, the materials being used are flammable and/or explosive. Luminaires for use in such conditions have very special characteristics[1]. Similarly, some laboratories may create conditions which are corrosive to the materials used in lighting equipment and the equipment selected must be chosen so as to withstand these conditions[1]. Other laboratories require lighting in areas where high temperatures are generated. Luminaires capable of surviving such temperatures will be needed and the choice of light sources may be restricted to those whose light output is largely independent of ambient temperature. Finally, it is necessary for the designer to remember that some materials used in laboratories are light sensitive. Where this is the case, the lighting installation and any form of day-lighting will need to be controlled.

OPTIONS FOR THE DESIGNER

Having described the aspects of the laboratory which the designer should consider, it is now necessary to examine the options which he has available to satisfy the requirements that have been identified. The options can be considered under four headings:

(1) Choice of light source.
(2) Choice of luminaire.
(3) Choice of layout.
(4) Choice of control system.

Choice of light source

Light sources are available in many different forms. Table 22.1 provides a summary of the properties of the main light sources likely to be used in laboratories in terms of their luminous efficacy, life, colour properties and ignition characteristics.

Table 22.1

Properties of the main light sources commonly used in laboratories

Lamp type	Range of luminous efficacy (lm/W)	Lamp life (h)	Colour properties	Typical time to full light output (min)	Typical time delay before re-ignition after switch-off (min)
Tungsten filament	8–18	1000–2000	Good	Immediate	Negligible
Tubular fluorescent	18–80	8000	Fair to excellent depending on lamp phosphor properties	Immediate	Negligible
High-pressure mercury discharge	35–85	8000	Fair to good depending on lamp phosphor properties and arc tube additives	5	10
High-pressure sodium discharge	50–110	11,000	Poor to fair depending on lamp pressure	7	1

The luminous efficacy of a lamp is expressed in units of lumens of light output per watt of power applied. Where a control circuit is necessary for the lamp to operate, the power used in this circuit is included. Control circuits are necessary for all the lamp types listed in Table 22.1 except the tungsten filament lamp. The luminous efficacy for each lamp is given in terms of a range because the luminous efficacy for the same lamp type can vary markedly with the wattage of the lamp and the colour properties it provides. It can be seen from Table 22.1 that there is a considerable overlap in luminous efficacy between different lamp types.

In Table 22.1, the life of each lamp type is given in terms of hours of use, but the events which determine the life of the tungsten filament lamp are different from those for other lamp types. The life of the tungsten filament lamp is determined by the failure of the filament. The lives of the other lamp types are usually limited by economic factors rather than total lamp failure because the light output deteriorates markedly before the lamp fails electrically. The lives given in Table 22.1 for the lamp types other than tungsten filament are those typically taken for the light output to fall to 30 per cent below the initial light output.

Whilst luminous efficacy and lamp life are important considerations because of their effect on the running cost of the lighting installation, there are other factors which influence the choice of light source. Where accurate colour matching is required, only tubular fluorescent lamps of a special class are suitable because these are the only lamps to meet the British Standard requirements[2]. In laboratories where accurate colour appraisal is required a wider selection of lamps is available, but the most efficient discharge lamp, the high-pressure sodium discharge, still cannot be used. It is only when the task to be done does not involve the judgement of colour that choice from the full range of light sources is possible.

Another aspect of light source properties which influences the choice made is run-up and restrike times. Not all light sources provide full light output immediately upon ignition; some require several minutes before a significant proportion of the total light output is available. Also, some light sources will not re-ignite immediately after being switched off. Neither of these aspects matter greatly if allowance is made for them in the overall design. Specifically, when light sources with long run-up or long restrike times are used, some form of back-up lighting system for use in emergencies will be necessary.

Choice of luminaire

Luminaires are available in many different forms. The properties of luminaires which affect their selection for use are utilisation factor in the laboratory under consideration, luminous intensity distribution, construction and appearance.

The utilisation factor of a regular array of luminaires of the same type in a specified laboratory is a measure of the efficiency of the installation in producing illuminance on the working plane and hence relates to the installation running costs. The luminous intensity distribution of the luminaire is also important as it affects the way in which three-dimensional objects appear and the extent to which discomfort glare is likely to be experienced by the occupants. The construction of the luminaires is a matter of suitability for the operating conditions.

All luminaires should comply with the relevant British Standard[3], but special operating conditions may require luminaires constructed in special ways to have suitable properties, for example, flameproof, jetproof, or splashproof luminaires. Appearance is not something that can be quantified easily but it is a factor which needs to be considered if a visually pleasing environment is to result.

Choice of layout

Having chosen the light source and the luminaire, the next choice open to the designer is the layout of the light source/luminaire package. Broadly there are three types of layout – uniform lighting, localised lighting and local lighting (Fig. 22.4). Uniform lighting is characterised by having the same illuminance everywhere over the working plane. By doing this, flexibility of use is ensured in that an illuminance appropriate for the tasks is provided everywhere, but it is also the most wasteful in that the task illuminance is provided in areas where there are no tasks to be done.

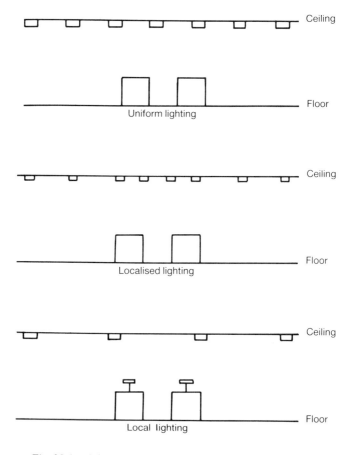

Fig. 22.4 – Schematic diagrams illustrating lighting layouts.

Localised lighting is characterised by having the task illuminance only in areas where tasks are to be performed, whilst areas where tasks are not performed are lit to a lower illuminance, usually about one-third of the task illuminance. Therefore, localised lighting is an inherently more efficient form of lighting than uniform lighting but the increase in efficiency is achieved at the expense of flexibility. If this form of lighting is used, the lighting installation may need to be altered when there is a change in work arrangements.

Theoretically local lighting is more efficient than localised lighting because the task illuminance is only provided in the immediate area of the task. Unfortunately, the light sources and luminaires commonly used for local lighting are relatively inefficient, thus the increase in efficiency may be illusory. However, the advantages of local lighting are not restricted to luminous efficiency. It can provide a degree of individual control of the amount and distribution of light. Where the task requires the discrimination of fine detail and particularly where there is some degree of obstruction to uniform lighting, then local lighting may be the most appropriate form to adopt. The choice between uniform, localised and local lighting is ultimately a matter of judgement, subject to the constraints imposed by particular features of the interior.

Choice of control system

One area of lighting in which rapid developments have been taking place recently is in control systems. Control systems in this context are means to switch and dim electric lighting installations in response to some from of signal. The signal can be provided by manual control, automatically at a fixed time or by the sensing of daylight availability or room occupancy. Such control systems have two main uses. They can be used to reduce running costs by ensuring that the electric lighting is only used when it is needed, during working hours, and also to convert a general lighting installation to a localised installation which can be changed to another work layout as desired. Control systems should always be considered when designing a lighting installation, no matter whether it is new or a refurbished one[4].

SUMMARY

Ideally, designing lighting for laboratories is a matter of tailoring the lighting installation to the requirements of the laboratory. The types of task to be done, the need to avoid discomfort and the conditions in which the lighting equipment will have to operate combine with financial and architectural considerations to determine the choices of light source, luminaire, layout and control system. Table 22.2 provides a checklist which can be used during the design of laboratory lighting to ensure that the right questions are asked.

It should be emphasised that the information given in Table 22.1 is indicative rather than definitive. During design, accurate, up-to-date information on source,

Table 22.2

Checklist for use by designers of laboratory lighting

Situation

What activities take place in the laboratory:
   do these activities require discrimination of detail, detection of movement, detection of form, colour appraisal or colour matching; are there any statutory requirements relevant to the lighting of the laboratory?

Where do these activities take place:
   is the location fixed, is the location likely to change with time?

Are the activities to be lit by the lighting installation being designed or is some form of special lighting associated with equipment to be used?

Do the environmental conditions in the laboratory limit the choice of equipment? Is there a hazardous environment present, i.e. one in which the operation of the lighting equipment might create a fire, explosion or other risk to safety?
Is the environment such as to cause corrosion of the materials used in the lighting equipment?
Are the environmental conditions such as to reduce the efficiency of the lighting, e.g. a high or low ambient temperature, dirty conditions?
Could the lighting equipment be used as part of a heating and ventilating system?
Are the materials handled in the laboratory sensitive to light?

What are the financial constraints on the lighting installation?

What are the architectural constraints on the lighting installation; has a planning grid been fixed; has the layout of large pieces of equipment been determined; are the room surface reflectances known; is there any daylight available; are the mounting positions for luminaires restricted?

Layout

What is the most appropriate form of lighting installation: uniform, localised or local?
How is electrical power to be supplied to the luminaires?
Where can the luminaries be mounted?
Is the proposed layout of luminaires compatible with other building services?
What degree of control of lighting is to be provided?
Should the installation be dimmable?
Are blinds required to exclude daylight?
Will the luminaires be easily accessible for maintenance?
What type of emergency lighting is required?

Selection of equipment

Does the light source have appropriate colour properties?
Is it optimised with respect to luminous efficacy, life and cost?
Are its ignition properties allowed for in the design?
Are replacement lamps readily available?

Is the luminiare efficient, suitable for the conditions in which it will have to operate and easily maintained?
Does the luminaire chosen comply with the relevant British Standards?

Is the photometric data readily available?

What type of control system is appropriate?
Is manual switching sufficient; can advantage be taken of available daylight and/or intermittent occupation to minimise energy consumption?
Are the light sources chosen suitable for use with the proposed control system?
Will the control system interfere with other instruments in the laboratory?

Assessment

Does the installation produce illuminances and glare indices consistent with the recommendations of the *Code for Interior Lighting*[5]?
Has care been taken to avoid veiling reflections from specularly reflecting surfaces?
Will the installation be free of flicker?
Does the installation match the financial and architectural constraints?
Will the appearance of the laboratory be satisfactory?

---

luminaire and control system characteristics should be sought from the manufacturers of such equipment. Further advice on the design of lighting installations is available [5,6]. If this advice is followed and care is taken in the analysis of the activities undertaken in the laboratory, then lighting which is both suitable and efficient can be produced.

REFERENCES

[1] *CIBS Lighting Guide: Lighting in hostile and hazardous environments,* The Chartered Institution of Building Services, London, 1983.
[2] BS 950: 1967, *Artificial Daylight for the Assessment of Colour,* British Standards Institution, London, 1968.
[3] BS 4533, *Luminaires,* British Standards Institution, London, 1971–1981.

[4] G. T. Depledge, Laboratory Lighting − Planning for Change, *CIBS National Lighting Conference 1982, Proceedings,* p. 13, Chartered Institution of Building Services, London, 1982.

[5] *Code for Interior Lighting,* Chartered Institution of Building Services, London, 1984.

[6] S. L. Lynes, *Handbook of Industrial Lighting,* Butterworths, London, 1981.

23

Low voltage electrical installations in laboratories

P. E. Donnachie
Property Services Agency, Croydon, Surrey.

INTRODUCTION

A high degree of flexibility is generally expected in modern laboratories to accommodate the ever-changing requirements of the user. This flexibility, or lack of it, determines to a large extent the uses to which a laboratory can be put and, in this respect, the services play a crucial part. Frequently the main features of the services required are not properly identified or adequately specified at the early planning stage. The result is that the objectives of flexibility are usually compromised, virtually ensuring that future changes in requirements will necessitate major work and cause considerable inconvenience and delays in the laboratory's work programme. In the case of electrical installations this picture is likely to improve in the future, but at a cost.

REGULATIONS GOVERNING LOW VOLTAGE INSTALLATION PRACTICE IN THE UNITED KINGDOM

The design and erection of electrical installations in the United Kingdom should be carried out in accordance with the Institution of Electrical Engineers (IEE) Wiring Regulations[1]. While these are not statutory regulations they have long been regarded as the national standard. The latest edition of these Regulations is based on international wiring rules[2]. Other countries are revising their national wiring regulations along similar lines. The result of this international influence has been to direct the IEE Wiring Regulations mainly at the professional engineer engaged in the design of electrical installations; that is, they are presented in the form of a national design guide. The latest Wiring Regulations are also more rigorous and comprehensive than previous editions and are equally applicable to large and complex industrial or commercial installations, and to domestic installations.

Primarily concerned with safety, especially from fire, electric shock and burns, the Regulations also impinge on operational aspects of installations. Although designers are given considerable freedom of choice they are required to ascertain all the relevant information relating to the purposes of an intended installation, the environment into which it will go, and who will use it. The object is to provide a completed installation that is safe and suitable for the conditions and usage 'throughout its intended life'.

PLANNING LABORATORY INSTALLATIONS

The electrical designer is faced with a number of problems in attempting to achieve a high degree of flexibility. Critical features that must be settled at an early stage in the planning of a laboratory, and upon which the detailed electrical design is based, include the type of distribution to be employed to provide supplies to benches and other working positions, the nature and extent of isolation and emergency switching facilities, and the shock protection measures to be incorporated. These features are discussed in some detail, in relation to the requirements of the Wiring Regulations.

An assessment of the types, numbers and distribution of electrical loads has to be made as soon as the client's purposes for a laboratory, both present and, as far as is possible, future, have been established. This information is then used to estimate the maximum demand (the peak value of the expected steady electrical load), as required by Regulation 311-1 of the Regulations. A decision must then be taken on the type of electrical distribution system to be employed within the laboratory. This can have a crucial effect on the working environment in respect of the ease with which future changes in layout or of use can be accommodated. The chosen system may also make a significant spatial and visual impact.

DISTRIBUTION AND CIRCUIT ARRANGEMENTS

In multi-storey laboaratory accommodation the main electrical distribution from the source of supply to each floor will resemble that for any other multi-storey accommodation, and will generally comprise one or more three-phase and neutral vertical busbar systems (risers) with take-off points on each floor. Regardless of whether the accommodation is single-storied or multi-storied, there are three main types of horizontal distribution system to choose from — floor, wall (peripheral), and overhead (suspended). The choice of system should be made only after careful consideration has been given to the detailed requirements of the Wiring Regulations and, indeed, to the relevant requirements of safety legislation covering places of work[3]. Section 314 of the Wiring Regulations deals with circuit arrangements and bears directly on the type of system most appropriate to a given situation, be it in a laboratory or elsewhere. The requirements of Regulation 314-1 in particular should be noted:

Every installation shall be divided into circuits as necessary to —
(i) avoid danger and minimise inconvenience in the event of a fault, and
(ii) facilitate safe operation, inspection, testing, and maintenance.

This covers much more than appears at first reading; but one fairly obvious aspect is that of the avoidance of overloading. This could be a serious problem where the load density is likely to increase in the future in a way that might be difficult to predict at the design stage.

The floor system of distribution is the least favoured of the three systems. To make adequate provisions for both present and future needs a grid of floor outlet locations is necessary. Many of these outlets may not be readily usable at any one time either through obstruction by benches and other furniture or because they coincide with circulation spaces. The system can often, therefore, be both restrictive and uneconomic. Floor outlets also give rise to trip hazards, and the ever-present likelihood that dust, solids or liquid spillage, including corrosive, volatile or contaminated substances, could enter the outlets and any underfloor trunking or ducting system associated with them, with possibly disastrous results. This type of system is therefore not considered to be a suitable first choice for laboratories of any sort.

The wall, or peripheral, system of distribution also has shortcomings. Measures must be taken to eliminate trailing cables between the walls and benches placed at right angles to them, by, for example, placing the cables in special in-floor ducts provided at each take-off point. The avoidance of trip hazards limits the arrangement of benches to configurations which create blind alleys. Wherever hazardous substances are in use there should always be two means of escape from the risk area and blind alleys present a real hazard in that they increase the risk of persons being trapped in the event of a fire or other dangerous occurrence within the laboratory. Another common problem with the peripheral system is overloading. This is most likely to arise in deep laboratories, where several benches may be butted end-to-end with their electrical supplies connected in a daisy-chain manner. If such a system is chosen and developed in this way the likelihood of overloading must be minimised by the provision of an adequate number of peripheral circuits of sufficient capacity. These should be coupled with clear, mandatory and enforced procedures for the connection of bench supplies, issued by the person responsible for the installation design (see the section 'Operating Procedures'). In shallow laboratories (measuring up to 7–8 m from the window wall) there should be little difficulty in avoiding overloading and the peripheral system may be preferable to an overhead system.

The overhead, or suspended, system is the preferred type where maximum flexibility is sought and in particular for deep laboratories. Exceptionally, as in chemistry laboratories where large glass structures may be present, a peripheral system may be more appropriate, irrespective of the laboratory depth, in order to avoid accidents which might otherwise occur from the movement of suspended cables or difficult access to electrical outlets. The overhead system can take several forms, the usual one being a series of suspended busbar or boom runs along the length or breadth of the laboratory. These runs are provided with tap-off points from which flexible droppers connect either directly to equipment or, perhaps more conveniently, to local distribution trunking systems containing outlets associated with individual benches or sets of benches. The potential for overloading of circuits is not entirely eliminated since it exists in practically all electrical installations. However, the system can in the main be traced by visual inspection and since daisy-chaining of bench supplies is not necessary, the installation can be more readily controlled.

The overhead system can have a major spatial and visual impact and integration of the system with other competing services, including the general laboratory lighting, can provide a pleasing and practically ideal result. Achieving integration is essentially a matter of proper coordination, with particular attention being paid to orientation, spacing and mounting height, to avoid clashes, shadow effects or problems of accessibility. Coordination can be aided considerably by the use of special booms that contain both electrical and other services, but it is necessary to check that their design and construction satisfy Regulation 525-11. Restrictions on the use of booms with regard to certain piped gases must also be observed.

ISOLATION AND EMERGENCY SWITCHING FACILITIES

There is often confusion, even in the minds of electrical installation designers, about the purposes of isolation and emergency switching; the Wiring Regulations help, by drawing a clear distinction between them. Isolation is the cutting off of an electrical installation, a circuit or an item of equipment from every source of electrical energy. It is normally an *off-load* operation that de-energises the live conductors of the isolated installation, circuit or equipment by an approved method which allows work to be safely carried out on, or with exposure to, the isolated parts, or for other similar purposes. Emergency switching is intended for the rapid cutting off of electrical energy to remove any hazard to persons, property, etc., that may occur unexpectedly. It is therefore, an *on-load* operation, requiring the provision of suitable means for carrying it out without placing the operator in danger or causing a further hazard.

Common practice is for the means of isolation to be located in a readily accessible position, outside the doors leading into the laboratory. Where devices provided for this purpose are isolators in the strictly correct sense (the international name is 'disconnectors') they are suitable for off-load operation only and must not be used to switch normal loads, or in emergency situations. Isolators should be available for use under no-load conditions by skilled persons only, as defined in the Wiring Regulations. The devices should be located in a cupboard or rendered unavailable for use by unskilled persons by some other means. Where the use of such devices is unrestricted, isolating switches (internationally known as 'switch disconnectors'), which are on-load devices, must be installed. These should not, however, be introduced to serve as both isolators and the sole means of emergency switching where there is a clear need for such a feature in the laboratory design.

The need for emergency switching (or tripping) and its arrangement are matters for agreement between the designer and the laboratory user or client. The wide range of activities, and their attendant hazards, likely to be carried out within one laboratory complex, coupled with the likely use of a considerable number of movable or portable electrical equipment items, frequently justifies the inclusion of emergency switching. The designer, when assessing the need for emergency switching should ask himself whether there is any valid reason why it should not be provided. The Wiring Regulations state clearly what form emergency switching devices should take, but do not provide detailed guidance on the extent of the facility. As a general rule, power circuits only are embraced, but not lighting. It may also be advisable, or perhaps essential, to exclude supplies to specific equipment from the emergency arrangements. For example, this could apply to fume cupboard extract systems, or to the supplies to experimental processes which might result in a catastrophe of some kind if the electrical supply is suddenly withdrawn. For these circuits it could be more appropriate to provide a 'controlled access' means of switching off the supplies, coupled with an audible and visual alarm system to give early warning if something goes wrong. Such switching cannot be described as emergency switching, which is provided for rapid (immediate) operation.

Isolation facilities usually present no difficulties where flexibility is required, whereas emergency switching may. One practical solution to overcome this difficulty is an area by area, or zonal, approach. It cannot be emphasised too strongly that devices for emergency switching must be readily accessible, preferably sited near each exit door and at any other agreed positions, clearly identified as to their function and, very important, periodically tested.

SHOCK PROTECTION MEASURES

The Wiring Regulations define two ways in which persons may receive an electric shock from an installation. These are *direct contact* and *indirect contact*. The first relates to making contact directly with normally energised (live) parts of electrical equipment, including any neutral conductor or terminal. The second relates to making contact with normally touchable metalwork of the installation (exposed conductive parts) made live under fault conditions, that is, making contact with live parts in an indirect way. Chapter 41 of the Regulations describes various methods of providing protection against both forms of contact.

Protection against direct and indirect contact can be readily achieved by the adoption of reduced voltage supplies but this concept has limited application in practice since much laboratory equipment is designed to operate on standard low voltage supplies. Even so, reduced voltages should be employed if at all possible in places where the risk of serious electric shock is greater than normal. School laboratories fall into this category since pupils may tamper with electrical equipment or disregard safety procedures. Young persons can also receive more severe electric shocks than adults, since smaller body mass results in a lower body resistance and the severity is related to the current passing through the body[4].

The Wiring Regulations describe means of providing protection separately against direct and indirect contact, for electrical installations operated at low voltage. Protection against direct contact is most often provided by insulating live parts and/or placing them inside enclosures. Special measures as described in Section 471 of the Regulations may be required, where, of necessity, there is some degree of exposure to live conductors. The special measures of protection against indirect contact include making the working location non-conducting or employing the technique of electrical separation. In many laboratories these measures cannot be employed and the most generally applied protection measure is bonding and earthing of exposed conductive parts and the metalwork of other services, coupled with automatic disconnection of the faulty part of the electrical installation without undue delay (Regulation 413-4). Disconnection may be achieved by operation of either the normal circuit overcurrent protective device(s) or a high-sensitivity (30 mA maximum) residual current device (RCD). The latter is more familiarly known as a current-operated earth-leakage circuit breaker. Sound, properly maintained, earthing arrangements, leading to efficient disconnection, are the key to minimising the shock risk in an installation. RCDs provide better protection than overcurrent devices, if they are properly maintained and regularly tested. RCDs should be employed in laboratories where plug-in equipment in large numbers is used or the configuration of the installation may be changed frequently. General laboratory lighting should not be included among the RCD-protected circuits and, as with some other features of the installation, this type of protection is best arranged on a zonal basis. This minimises the inconvenience of RCD operation and makes it easier to test the devices.

SELECTION OF EQUIPMENT

The provision of a laboratory electrical installation that is safe and suitable for the conditions and usage throughout its intended life can be realised only by the selection of appropriate equipment and materials. This implies the use of tough, durable, high-quality components, conforming to British or other internationally recognised Standards, that are able to withstand the rigours of frequent changes in working arrangements. Cheaper, less robust, electrical equipment often fails at the worst possible moment and its use is likely to lead to unnecessary breakdowns and higher maintenance costs.

A laboratory is perhaps best treated as an industrial area when it comes to the selection of equipment. On this basis socket outlets, plugs and other accessories for the connection of power-consuming equipment would normally be of the relevant British Standard[5], or other heavy-duty type, the domestic type being provided for housekeeping purposes only. Non-fused accessories, including those conforming with the British Standard, have serious repercussions for the design of final circuits and the overcurrent protection of equipment and these matters merit special attention by the designer. On the other hand, the advantages of colour coding and non-interchangeability that British Standard socket outlets and plugs offer can also be of considerable benefit in many laboratories where mixed supplies of different types and voltages are common.

Overcurrent protection of final circuits in laboratories is frequently provided by miniature circuit breakers (MCBs), in preference to fuses, because of the attraction of their quick reset facility. The MCBs are often also used, as a convenient dual-purpose device, for the functional control of circuits or equipment. The designer should not include MCBs on this basis without satisfying himself that the proposed arrangements comply with the requirements of the Wiring Regulations for functional switching. These include safe access to operate (Regulations 13-16 and 513-1) and clear identification of purpose (Regulation 514-1). Where MCBs are liable to be operated frequently, those selected must be appropriate to the duties they are expected to perform.

OPERATING PROCEDURES

All laboratories should have a proper instruction manual which sets out the correct procedures for using the electrical installation; any limitations imposed by the design must be identified. The manual should also contain clear instructions on the use of any emergency facilities. The designer should be consulted if, at any time, there is a requirement to depart from the terms of reference dictated by the electrical design.

REFERENCES

[1] *Regulations for Electrical Installations,* 15th Edition, Institution of Electrical Engineers, London, 1981 and its Amendments dated January 1983 and May 1984.

[2] *Electrical Installations of Buildings,* Publication 364, International Electrotechnical Commission, Geneva, 1972 and subsequent parts.

[3] The Health & Safety at Work, etc. Act 1974, Chapter 37, Section 6, HMSO, London.

[4] *Effects of current passing through the human body,* Publication 479, International Electrotechnical Commission, Geneva, 1974.
[5] BS 4343: 1968, *Industrial Plugs, Socket Outlets and Couplers for AC and DC Supplies,* British Standards Institution, London, 1980.

24

Electrical safety in laboratories

K. Oldham Smith
Electrical Safety Consultant

This chapter deals with the legal requirements for electrical safety. It seeks to identify the electrical hazards and the appropriate safety precautions which are available in any kind of test facility or laboratory.

LEGAL REQUIREMENTS

Before the inception of the Health and Safety at Work, etc. Act 1974 (HSW Act), the only appropriate legislation was the Electricity (Factories Act) Special Regulations 1908 and 1944. This legislation, which is still in force, adequately covered the electrical safety requirements but was only applicable to premises covered by the Factories Act. Research laboratories, school and college laboratories, for example, were not covered. Whereas the application of the Factories Act was dependent on the status of the premises, the HSW Act is applicable to all persons at work whether they are employed or self-employed. It also affords protection to others in the vicinity of the work site. This legislation is enforced by the several Inspectorates within the Health and Safety Executive.

The HSW Act does not include any specific electrical safety requirements but it does set out the general principles. For example, in Section 2 (2) (a) the requirement for 'the provision and maintenance of plant and systems of work that are, so far as is reasonably practicable, safe and without risks to health'. To determine the specific requirements and what is reasonably practicable, reference must be made to any other relevant legislation such as the Electricity Regulations and other non-statutory safety rules and codes of practice such as the Institution of Electrical Engineers Wiring Regulations and appropriate British Standard Specifications. To aid testing and research work, Exemption 4 of the Electricity Regulations provides exemption for

'any process or apparatus used exclusively . . . for testing or research

purposes provided such process be so worked and such apparatus so constructed and protected and such special precautions taken as may be
necessary to prevent danger.'

Devising 'special precautions' can be onerous. It is advisable therefore to adhere
to the requirements of the Regulations as far as possible and avoid the need to
devise these precautions.

In Section 6 of the HSW Act there is an onerous requirement which is
imposed on designers, manufacturers, importers and suppliers of articles for use
at work who have to ensure the article has been designed and constructed so that
it will be safe when properly used. They also have to test it to prove its safety
and provide information about it so that the user can install and operate it safely.
This requirement is also applicable to those who design rigs for experimental
work. Such facilities, although temporary, have to be safe. Paragraph 2, of
Section 6, requires the designer or manufacturer to carry out any necessary
research to eliminate any design faults so that the article will be safe in use. To
do this, the designer/manufacturer needs to familiarise himself with laboratory
practices and visualise the various possible usages of his apparatus. As he has to
make available information about the safe utilisation of the equipment it is
advisable for him to indicate in the instruction manual any safety limitations on
its use.

The effect, of the HSW Act, is to impose duties of care on everyone concerned with the work environment whether they are the owners of the property,
the self-employed, employers, managers, employees or suppliers of the plant,
installations and equipment.

HAZARDS

As the electromagnetic spectrum extends from cosmic rays to a dull thud
there is plenty of scope for a wide range of potential electrical hazards. To
limit the field and avoid discussing hazards dealt with elsewhere, such as ionising
radiations, these remarks are confined to the hazards of shock, burn, fire and
explosion.

Shock

Electric shock is experienced when a current, above the threshold limit of
perception, passes through the body. It can be dangerous if it traverses any of
the vital organs. Table 24.1 relates the current to the shock phenomena experienced for alternating current at about mains frequency. These figures assume the
current traverses the vital organs, i.e. heart and lungs, and results from the usual
hand-to-hand or hand-to-foot shocks, and that the victim is in normal health.

Smooth DC, as from a battery, produces shocks which are less dangerous
because the shock sensation is not continuous. Shocks are experienced only
during current changes and so the victim gets two shocks, one on initial contact

Table 24.1

Shock phenomenon	Current (RMS value, mA)
Threshold of perception	1.00 to 2.00
Mild to painful shock	2.00 to 10.00
Inability to 'let go', i.e. some paralysis of the motor nerves	10.00 to 20.00
Respiration ceases and/or cardiac arrest causing loss of consciousness	20.00 to 50.00
Ventricular fibrillation	over 50.00

when the body current rises from zero to a steady state value and another when contact is broken and the current declines to zero.

The body impedance, between the shock contacts, is measured in some thousands of ohms and depends on a number of variable factors such as area of contact, skin dryness and applied voltage. The impedance varies inversely with the voltage. Under normal dry conditions, shocks are unlikely at safety extra low voltage, i.e. not exceeding 50 V AC. Electric shock damage is time dependent. The International Electrotechnical Commission has published time/current curves indicating the duration of tolerable currents[1].

The shock sensation is frequency dependent. Dalziel and his co-workers carried out experiments on human volunteers and measured the maximum 'let go' current at various frequencies[2]. His results indicate that the United Kingdom has chosen the worst possible frequency for its electricity supply as the 'let go' current is at a minimum value at 50 Hz. At frequencies of about 1 MHz and above the shock sensation disappears.

Burns
Electrical burns occur from contact with a power arc, from the passage of current through the body and from radiation. The intense heat of the electric power arc produces very severe burns which are often deep-seated and which take a long time to heal. Current burns often occur when the victim experiences an electric shock. They are apparent at the points of contact if the current density is sufficient. Large area contacts with lower current densities may not show surface burning but there could be damage to internal tissue. Radiation includes burns from infrared, ultraviolet, coherent light and microwave energy produced by, for example, hot wire elements, arc welding, lasers and magnetrons.

Fire and explosion
Electrically caused fires and explosions are not as common as shock and burn accidents and are rather less likely to injure people, but they cause quite a lot of property damage. In large electrical laboratories, there may be oil-immersed

equipment such as transformers and switchgear which may explode on the occurrence of an internal short circuit. Elsewhere, faults and gross overloading of cables and other electrical equipment, can cause arcing or overheating and lead to a fire. Flammable atmospheres can be ignited by an incendive spark if non-explosion-protected equipment has been used or if the apparatus temperature reaches the ignition temperature of the flammable substance or gas. Dust explosions can be initiated from an electric spark if the apparatus is not dust tight.

SAFETY PRECAUTIONS

It is impossible to cover the very wide field of safety precautions adequately in this chapter. The various safety measures may be listed under broad headings.

(a) Provision of a safe electrical installation and apparatus.
(b) Protection against excess current and electric shock.
(c) Interlocking.
(d) Explosion protected apparatus.
(e) Non-ionising radiation protection.
(f) Maintenance.
(g) Staff training.

Each will be discussed briefly.

The electrical installation

The electrical installation should comply with the Electricity Regulations, the non-statutory Institution of Electrical Engineer's *Regulations for Electrical Installations* and any relevant codes of practice. It must be compatible with the environmental conditions. In corrosive atmospheres, for example, non-metallic conduit and/or trunking should be used with plastics luminaires, switches and socket outlets sealed or enclosed to exclude the corrosive gas as far as possible. In extreme cases, the installation may have to be purged with a flow of clean air inside to prevent the ingress of the corrosive gas.

Where a variety of DC and AC supplies are used, non-interchangeable plugs and socket outlets are necessary. The socket locations should be decided so as to avoid long trailing leads and there should be an adequate number to avoid the use of adapators.

Electrical apparatus should not be located in places where there is an increased risk of electric shock, such as in wet and/or confined conductive spaces, unless it operates at safety extra low voltage, or other special precautions are taken.

The maker, of the apparatus, has a duty to supply only safe equipment. If there is any doubt, an assurance should be sought that the apparatus meets the requirements of the HSW Act, Section 6.

Excess current and shock protection

Excess current protection covers both overload and fault protection. The protective device whether fuse or circuit breaker, must be capable of interrupting the prospective fault current in the event of a short circuit. The prospective fault level can be ascertained from the supply authority or by measurement.

Electric shock from fixed apparatus is less likely provided that it is properly installed and maintained. Portable apparatus is more of a shock risk. The flexible cables and their terminations produce more accidents than apparatus faults. Figure 24.1 illustrates a typical Class I apparatus shock accident. The cord grip fails, the protective conductor is pulled out of its terminal and touches the phase terminal or fuse assembly causing the unearthed apparatus to become 'live' at the supply voltage. Anyone handling the apparatus when in contact with anything earthed, would get a mains voltage shock. The remedy is to use plugs with adequate internal barriers and maintain the cord grip. Alternatively, Class II apparatus which does not have a protective conductor and is unearthed should be used. It is better still to use low voltage apparatus, operating at 110 V and fed from a transformer with the centre point of its secondary winding earthed (Fig. 24.2). This reduces the prospective shock voltage to earth to 55 V which is comparatively harmless. If, however, mains voltage apparatus has to be used, it is

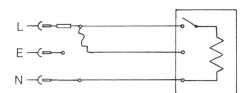

Fig. 24.1 – Shock hazard from 'live' unearthed apparatus.

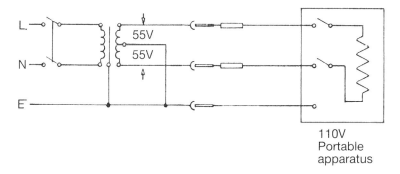

Fig. 24.2 – 110 V centre-tapped to earth system.

possible to ensure the integrity of the earthing system by circulating current earth monitoring (Fig. 24.3). Any open circuit, in the protective conductor, causes the contactor to open and cut off the supply to the apparatus.

Another system employs a residual current device (Fig. 24.4), which disconnects the supply, in the event of an earth leakage exceeding about 25 mA, in not more than 30 ms. The shock duration is too short to cause physiological damage.

In electrical and electronics testing laboratories, it is sometimes possible to create an earth-free environment which avoids the danger of shocks between anything 'live' and earth. Where, however, the metalwork of the apparatus has to be earthed, it is sometimes possible to insert a high impedance in the connection to earth which limits the possible shock current to earth, from either pole of the supply, to a harmless value of about 2 mA or less. The system can be arranged to disconnect the supply in the event of earth leakage (Fig. 24.5).

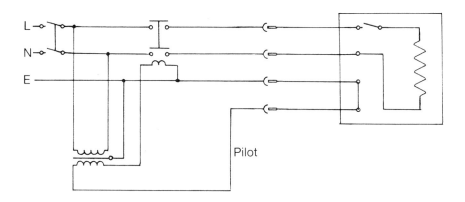

Fig. 24.3 – Circulating current earth monitoring.

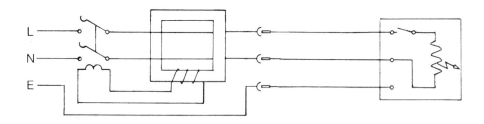

Fig. 24.4 – Earth leakage circuit breaker protection.

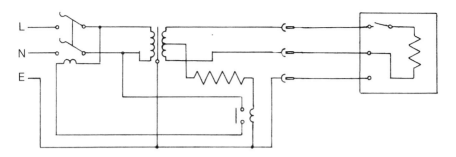

Fig. 24.5 – Protection by high impedance supply earthing.

Interlocking
One example occurs in high voltage and high power electrical laboratories where there is an enhanced risk of electric shock and danger from failures of apparatus with explosive violence. It is usual to segregate the equipment under test in an enclosure. The access gate is then interlocked with the supply switch, by means of a trapped key system, so that the gate can only be opened with the supply *off*.

Explosion-protected apparatus
To prevent ignition of the flammable atmosphere the following measures are normally employed.

(1) Flameproof apparatus type 'd' – the carcase is strong enough to withstand an internal explosion. It is constructed with wide flanges or labyrinthine joints between the several parts. These act as flame traps and prevent ignition of the outside flammable gas from an internal explosion. Figure 24.6 illustrates a typical installation of type 'd' apparatus.

(2) Increased safety apparatus type 'e' – similar to ordinary industrial equipment in appearance but specially designed to avoid the production of an incendive spark or to have host spots.

(3) Non-sparking apparatus type 'N' – suitable only for areas where flammable materials are processed, handled or stored but where the controls are such that a flammable atmosphere is only likely to occur under abnormal conditions (Zone 2 areas).

(4) Intrinsically safe apparatus type 'i' – this is low power, specially designed apparatus with circuitry which avoids the production of an incendive spark even under fault conditions.

(5) Pressurised or purged apparatus type 'p' – an inert gas or air is used to pressurise or purge the apparatus to prevent the ingress of the flammable atmosphere. Examples are shown in Fig. 24.7.

(6) Other measures are referred to in the various parts of BS 5345, *Code of practice for the selection, installation and maintenance of electrical apparatus for use in potentially explosive atmospheres (other than mining applications or explosives processing and manufacture).* Types 'o' and 'q' — oil-immersed and sand-filled apparatus — are not normally used in the United Kingdom.

Fig. 24.6 — A Thorn-EMI flameproof lighting installation in an ink maker's laboratory. (Reproduced with permission of Thorn-EMI Lighting Ltd.)

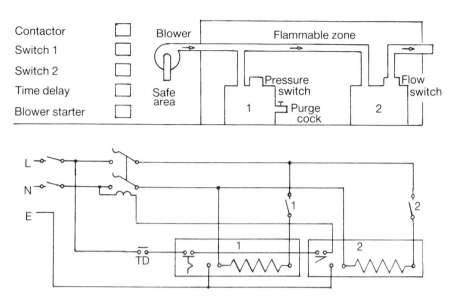

Fig. 24.7 – Protection of hazardous area apparatus by pressuring and purging. *Note.* The control gear located in the safe area is non-flameproof. The pressure and flow switches which are in the flammable zone must be of the explosion-protected type.

Non-ionising radiation protection

Microwave apparatus, such as ovens, is becoming increasingly used. Special techniques are necessary to prevent energy leakage from wave guides and resonant chambers. These are combined with the use of a detecting instrument, a power density meter, scaled in mW/cm^2. The leakage is contained by keeping joint faces clean, and using wave traps and energy-absorbent materials. The acceptable leakage limit is 10 mW/cm^2.

Maintenance

No installation will remain safe for long unless it is properly maintained. A scheme should be formally introduced and the necessary inspections and tests carried out at predetermined intervals and any repairs effected. Proper records are essential to ensure complete coverage and to determine any alterations in the maintenance scheme which the records indicate are needed.

Staff training

Safety training for staff is a statutory requirement. It should be tailored to suit the individual needs of the various grades of staff and the work they do. In the author's view, the sort of general safety training which may or may not feature in an industry training board course is not adequate and should be supplemented by specific training particular to the laboratory concerned. The training should include a reference to the company's safety policy and their electrical safety rules, if any. In the case of electrical and electronic laboratories where there is an enhanced risk of electric shock, artificial resuscitation should also be included.

REFERENCES

[1] *Effects of current passing through the human body,* Publication 479, International Electrotechnical Commission, Geneva, 1974.
[2] C. F. Dalziel, E. Ogden and C. E. Abbot, *Trans. Am. Inst. Elec. Eng.,* 1943, **62**, 745.

25

Special gases installations in laboratories

R. A. J. Kinsey
British Oxygen Company, London†

There are an increasing number of laboratories using special gases and the more hazardous types of these gases are being widely used in them. Many of these gases are extremely toxic, having very low Threshold Limit Values (TLV). Arsine and hydrogen selenide are two examples, with TLVs of 0.05 ppm[1]. Others are highly flammable – some, such as silane, being pyrophoric. The potential operational risks of these laboratories have risen and it is vitally important to the employer and the employee that steps are taken in the design of the facility to minimise these risks, and that correct operating procedures are devised and practised.

GAS SUPPLY

In most cases there is a range of different gases in continual use and in this situation, the place for all the cylinders to be stored is not in the laboratory where they are being used. If it is desired to use one new hazardous gas for a short period, the location of the cylinder in the laboratory may be permitted if there are adequate safeguards. These are that a minimum size of cylinder is located in a gas cabinet, and all who work in the vicinity, or who might be required to assist in an emergency, are made aware of the presence of this particular gas.

With old buildings, problems may arise, due to the design of the structure and the location of the laboratories, in siting the cylinder store and running the supply lines from it. It is disappointing, however, to find the same problems arising in new construction, because the specialist was not consulted before irrevocable decisions had been taken. The cylinder store should be external to the main building where it can be neutralised, instead of adding to the difficulties, if a fire should occur in the building.

† Now with Bluemain Ltd, Dorking, Surrey

Toxic, flammable, oxidising and pyrophoric gases should be mutually separated from each other in the cylinder store. Imperforate walls can be used to achieve this separation, provided they do not interfere significantly with the ventilation of the store. The store should, of course, have high and low level natural or forced ventilation, be constructed of light non-combustible material and have lightly fastened doors, which open outwards, in the external walls. Warning notices should be posted and breathing equipment should be available. Nothing should be stored in the vicinity which, if itself hazardous, could imperil the safety of the cylinder store.

PIPING

Most permanent gas cylinders have a maximum pressure of 2500 psig, although the suppliers are now moving towards a maximum of 3000 psig. It is obviously undesirable to pipe gases at this pressure through buildings, when the operating pressure in the laboratory is only 50 psig or less. Pressures should be reduced by installing regulators in the cylinder store and steps should be taken to limit the flow should the pipe become damaged. It is possible to install an ordinary excess flow valve, but flow limit, shut-off valves are now available which can be pre-set at the factory to operate at any flow rate between 100 and 50,000 cc/min. An alternative form of protection is to install an automatic shut-off valve at this point, either electrically or pneumatically operated. The advantages of this method are that the system can be shut down automatically, either because a number of conditions are outside the specification, or by a decision on the part of the operator.

The route of the piping through the building, if run in ducts, should comply with the Code of Practice which deals with ducts for building services[2]. It is sometimes difficult to comply with this code, and is impossible in the spaces behind laboratory benches where services are traditionally run.

Wherever possible, pipelines carrying extremely hazardous gases should be run outside the building, entering the building as close as possible to their point of use. Where pipes have to be run inside buildings, they should be run on the surface and, if there are many of them, neatly grouped on a cable tray. Pipework for pyrophoric gases, however, must be separated from other pipework so that in the event of a reasonably foreseeable leakage, any resultant flame will not impinge upon and damage other piping, etc. Such gas lines should never be run in ducts. Periodic leak tests are carried out more easily on pipes where the run is on the surface and such piping is more likely to be tested if not hidden in ducts. It is sometimes proposed that pipes carrying flammable gases are run in unventilated false ceilings; this should never be sanctioned. Even when the pipelines are run below the ceiling the void above should be ventilated, particularly where extremely light, flammable gases such as hydrogen are concerned. The gas could accumulate steadily over a long period, until there is sufficient quantity to case a disastrous explosion.

The materials of construction of the pipeline and all its components exposed to the gas must be compatible with the gas being used. Where extremely high purity gases are in use and the purity must be maintained, most of the piping should be of stainless steel to minimise pick-up. Where oxygen is concerned care must be exercised to ensure that the pressure and flow limitations are not exceeded if spontaneous combustion is to be avoided. Copper piping is the safest material for oxygen and should be used wherever possible.

Welded pipeline joints are generally held to be better than compression fittings, but it is not always possible to eliminate compression fittings completely from the system. Modern compression fittings have improved considerably but not all types are suitable and it is essential that they are installed strictly in accordance with the manufacturer's instructions.

An additional safety factor is that most fittings are designed for a working pressure of 2000 psig or more, but operating pressures are often no more than 50 psig. A high factor of safety can be achieved if the system is tested to the maximum working pressure of the weakest component. If gas leak testing is carried out, care must be taken to follow the requirements of the relevant British Standard [3].

Sometimes final pressure control is necessary at the point of use. Line regulators can be fitted, which have a bonnet relief valve opening to provide against diaphragm failure. Relief valves may also be fitted. In the case of flammable and toxic gases, each of any connections of this type should be piped away and vented to a safe level.

Piping must be identified by labels which state clearly the gas being carried. Although some colour coding can be carried out, no British Standard is sufficiently comprehensive to cover all gases likely to be used in laboratories. All pipelines carrying flammable gases must be effectively earthed and separated from electrical systems by at least 50 mm.

COMPONENTS

Many of the gases now used in laboratories have very low TLVs and some designers attempt to minimise the possibility of leakage by resorting to bellows or diaphragm type valves. This can be a mistake if the gas is highly corrosive since the unexpected presence of moisture can result in failure of the thin bellows after less than a day's operation. In these situations, it is preferable to use a more robust type of valve with packed glands that can suffer a degree of corrosion.

After every precaution has been taken to select suitable components and to run the piping in a manner least likely to cause danger, then steps should be taken to deal with the unexpected. Consideration should be given to the use of solenoid valves sited close to the point of use and activated by emergency buttons or by flow or pressure switches when conditions deviate from specification. The implications of a failure in the electricity supply or the ventilation system must also be taken into account. Operator training for both normal operation and abnormal situations is extremely important.

PURGING

Purging is a process which should receive attention in the design stage if safe and satisfactory operation of the system is to be achieved.

Purging is necessary to:

— establish as quickly as possible and to maintain a high purity gas stream when using high purity gases;
— prevent toxic or flammable gases entering the working area;
— avoid possibly corrosive conditions arising from acid-forming gases coming into contact with air or moisture;
— reduce the waste of expensive gases; and
— remove contaminants, which might affect the process, from dead ends, such as Bourdon tubes.

Frequently occurring situations where purging is required are when attaching a new cylinder to the delivery system and putting it into service, and disconnecting a cylinder from the system and isolating the system.

Prior to disconnecting a cylinder from the manifold, it is necessary to exhaust safely the process gas contained in the manifold and to purge the manifold with an inert gas until the concentration of the process gas is reduced to an accepted level. Before putting a new cylinder into service, the manifold must be purged with an inert gas to remove any air remaining between the cylinder valve and the first isolation valve of the system. It is also necessary to ensure that the concentration of any remaining impurities is not detrimental to the process.

There are two basic methods of purging — flushing, and dilution.

Flushing

In purging by flush flow, one end of the system is connected to an inert gas supply and the other to a vent line. The process gas is displaced by a continuous flow through the system, for a length of time which can only be determined by experience. Removal takes place through a combination of turbulent mixing and entrainment, but the effectiveness of this operation will be affected by any dead-end side capacity such as the Bourdon tubes of gauges. One flush flow purge through a system might reduce the gas concentration to about 200 ppm, but in a similar system with side capacity, the concentration is more likely to be of the order of 800 ppm.

Dilution

The more efficient way of purging is by dilution whereby the system is successively vented to a low pressure (atmospheric or lower), pressurised with purge gas to several atmospheres, and then vented again to the low pressure level. The cycle is repeated several times. Each cycle reduces the concentration of the process gas by a factor directly related to the ratio of vent pressure to purge gas pressure. The concentration is reduced by injecting into the system a greater volume of purge gas than the initial volume of process gas. If the pressure in the system is increased from atmospheric to 60 psig by the addition of purge gas, the system will contain one volume of process gas and four volumes of purge gas and the concentration of the process gas will be reduced to 0.20 of its initial value. The ratio of the vent absolute pressure divided by the absolute pressure of the purge gas is the dilution factor. Thus at the end of n cycles, the concentration will be reduced to $0.20n$ of its initial value. Increasing the pressure of the purge gas and/or venting at a lower pressure will obviously increase the dilution factor. Thus, if the purge pressure is at 60 psig and the vent pressure reduced to 370 mm, then the dilution factor is about 0.1.

An impurity level of about 2 to 3 ppm can usually be achieved in about ten cycles provided that time is allowed in each cycle for conditions to stabilise so that the whole volume is purged.

Push-button type valves are now available on the market which make the purging of systems very simple and which cannot be left in the wrong position.

LEAKAGE

Reliance on the operator's nose for the first warning that something is wrong is an unsatisfactory method for detecting leakage. A separate instrument to monitor each individual gas is normally not necessary as some types can monitor a group of gases. When the gases to be used are known, consultation with instrument manufacturers will usually result in a reduction of the number of monitors required.

SUMMARY

When constructing special gas installations for laboratories, it must be ensured that:

(a) the cylinder store is properly designed and sited;
(b) the piping system is properly designed and installed;
(c) there are emergency systems should things go wrong;
(d) personnel are instructed properly in the operation of the system, particularly purging; and
(e) instruments are installed to give warning of leakage.

REFERENCES

[1] Health and Safety Executive, Guidance Note EH15, *Threshold Limit Values for 1980,* HMSO, London, 1981.

[2] CP 413: 1973, *Ducts for Building Services,* British Standards Institution, London, 1980.

[3] BS 3351: 1971, *Piping Systems for Petroleum Refineries and Petrochemical Plants,* British Standards Institution, London, 1981.

26

Safety aspects of special gases in laboratories

H. Spencer
Smith, Kline and French Research Ltd

The definition of the term 'special gas' as used in this chapter is any gas which is normally supplied under pressure in a cylinder. The only gases used in laboratories and not included in this term are the mains supply of natural gas, compressed air from an on-line compressor and, if it is regarded as a gas, live steam.

SUPPLY SYSTEMS

The method of supplying a special gas to the ultimate user requires a major policy decision. There are, at least in concept, two alternatives. Either each individual scientist or laboratory has, under local control, cylinders of each required gas, or the cylinders are located in a central point with a pipeline running through the building to deliver the gas to each laboratory.

There is no firm rule to guide the user as to which of these alternatives is the best one. This chapter examines some of the problems that have to be considered in reaching a decision.

Although dwelling on the problems it is not suggested that the problems are insoluble and many have been overcome, as is evident from visits to various installations around the country. The one problem that is never solved is that which was never considered. It is the intention of this chapter to provide not a checklist but an insight into the thought that might be given to the provision of supplies of special gases.

INDIVIDUAL SUPPLIES

When cylinders are supplied to individual workers or laboratories there are a number of problems that can arise.

The first hazard that has to be considered is fire. A cylinder of gas represents a closed pressure vessel and two things happen when the temperature increases. Firstly, the pressure of the contents increases and, secondly, the strength of the cylinder itself decreases and at some point the cylinder bursts. This was spectacularly demonstrated when, in 1976, there was a fire at a depot of a supplier of bottled gases at Letchworth in Hertfordshire. Explosions of bursting cylinders were clearly heard five to six miles away. Bursting is a danger to anyone attempting to fight a fire where cylinders of compressed gas are stored. In the United States of America the rules of the Interstate Commerce Commission require that gas cylinders are fitted with a safety device to prevent over-pressurisation. In the United Kingdom such devices are the exception rather than the rule, although they are always provided on cylinders of acetylene, liquefied petroleum gas, and carbon dioxide provided for automatic fire extinguishing systems.

This is not the only problem that has to be considered. The pressure in a gas cylinder can be between 200 psig (e.g. for hydrogen sulphide) and 2000 psig (e.g. for nitrogen) and therefore a pressure regulator must be used. The standard pressure regulator contains a thin diaphragm, usually of stainless steel, which can develop a leak after being used for some time. Such leaks are comparatively slow but fast enough to discharge the entire contents of a cylinder overnight. The results of this type of failure can be catastrophic.

An investigation into an accident involving propane-fuelled space heaters is contained in a Health and Safety Executive Report[1]. These heaters, which were left burning, went out during the night due probably to a lack of oxygen. The cylinders continued to discharge propane and when the building was opened by workmen the following morning there was a violent explosion resulting in considerable damage. This accident, although not caused by leakage through a worn diaphragm emphasises that there is enough gas in an average cylinder to form an explosive atmosphere in a building of substantial volume. Thus leaking cylinders can cause fire and they can also cause poisoning. Some of the gases supplied in cylinders are not flammable but are highly toxic and a leak can cause a building of substantial volume to fill with a lethal atmosphere. Even leaking oxygen is not without its hazards as was demonstrated by the fire which occurred on board HMS *Glasgow*[2].

Another problem with keeping cylinders inside the building is that they are bulky and heavy. Simply finding a home for them presents problems and they have to be moved continually within confined spaces. This problem is illustrated by an incident which occurred in the United States of America some years ago[3]. A number of cylinders, which were part of an automatic fire protection system, had to be moved to allow painters to work behind them. One of the cylinders was dropped and the valve damaged. The cylinder travelled along the corridor, struck a scaffold, knocking off a painter, who broke his leg in the fall, struck a wall and removed a number of concrete blocks. An electrician was chased 20 m down a corridor and the cylinder eventually came to rest in the refuse collection area some 50 m from where it started.

This incident illustrates three problems. Gas cylinders occupy space and get in the way, they are heavy and difficult to handle, and they contain a lot of stored energy.

PIPED SUPPLIES

If the alternative solution of locating all cylinders at a central point and supplying each laboratory by means of a pipeline through the building is chosen, the problems are different but no less important.

A change to a central supply involves not simply engineering or design changes to the building but also an organisational change for the unit which eventually is to work in the building. Arrangements have to be made for the cylinders to be monitored and changed when they are empty. This function is usually the responsibility of the stores or site services rather than the laboratory scientists and therein lie the seeds for further problems.

One problem is that the person who changes the cylinders has no way of knowing whether every valve on the pipeline is closed when a new supply is connected. In 1979 a school in Clwyd suffered considerable damage from just this cause[4]. The rural school was not connected to a mains gas supply and the laboratories were supplied with propane from cylinders housed outside the building. The cylinders became empty just before a half-term break, during which the cylinders were changed. The men who changed them had neither the responsibility nor the access to check that every outlet tap was closed. In fact, they were not and the contents of the new cylinders were discharged into the laboratory through an open tap. Something in the closed school ignited the explosive gas/air mixture and considerable damage was caused.

Another problem likely to arise when cylinders in a central bank are changed by non-scientists is the connection of the wrong cylinder to a pipeline. Kletz has published some figures on human reliability[5] from which it can be concluded that if it is physically possible to interchange cylinders then it is only a question of time before such an error occurs. This has been recognised in hospitals where the result of supplying nitrous oxide instead of oxygen would be catastrophic and the Department of Health and Social Security has issued instructions on how these two gases should be supplied[6].

In a conventional laboratory situation another potentially disastrous interchange is supplying oxygen for nitrogen. The main use of nitrogen in the laboratory will be to provide an inert atmosphere for reagents which are oxygen-sensitive. Oxygen supplied as nitrogen will certainly mean the loss of the experiment but may also result in an explosion and fire. If neither of these effects occur then it is likely that another problem has already highlighted the substitution. Most oils and greases can ignite spontaneously in a pure oxygen atmosphere and therefore all pipes, valves, etc. in an oxygen system would be rigorously degreased. Such precautions would not have been taken for a nitrogen supply system and therefore substitution with oxygen may result in ignition.

A further problem associated with the changeover of cylinders is the possibility that pressure in the pipeline may drop below the normal operating pressure. This could result in the pipeline pressure being lower than the systems connected to it and chemical reagents may be sucked back down the pipework. This could lead to corrosion and leaks, or to chemicals soldifying in control valves which in consequence may seat badly. It is also possible that when the supply is re-established there may be a violently exothermic reaction leading to extra high pressure in the line.

LOSS CONTROL

The problems and incidents described so far have been concerned with the classic problems of safety, that is the prevention of fires and explosions, and exposure to toxic gases. Safety is, however, also concerned with the prevention of any kind of accidental loss. As an example a small amount of air introduced into a pipeline during the changeover of a cylinder can represent an unacceptable loss. In a typical situation, super high purity helium, guaranteed not to contain more than 10 ppm of impurities including nitrogen, and costing £1,500 per cylinder, may be used. To allow a small amount of air into the system when the cylinder is changed would result in an unacceptable oxygen contamination. Removal of this contamination might consume gas worth hundreds of pounds.

MAINTENANCE

Any equipment used with special gas supplies, whether part of an installed pipework system or part of an individual system, needs regular maintenance. It may be necessary also, from time to time, to modify an installed pipework system. Special safety problems arise when any work has to be undertaken on installed pipeline distribution systems. Such an operation presents hazards for two groups of people; the laboratory user who has planned work on the basis of a continuing supply of gas, and the maintenance team who have planned their programme on the basis that the line is isolated and purged. Safety depends upon mutual agreement and clear understanding between three groups, the laboratory user, the maintenance engineer and the person who controls changing, removal or isolation of the cylinders. A further requirement is a clear identification of the services as they run through the building.

PRESSURE

Different users will have different requirements for the pressure at which a gas is to be supplied. The temptation will be to adopt the highest pressure needed by users as the supply pressure and this may not be a good solution.

The joints of standard laboratory glassware can be forced apart and leak at pressures as low as 0.2 psig. If wired together to prevent them being forced apart then the glassware would withstand pressures of 40 to 50 psig before becoming increasingly likely to fracture due to excess pressure. Laboratories intending to use special gases in standard laboratory glassware cannot use, with safety, gas supplies at 80 to 100 psig, although other users may have a need for such pressures. Clearly, therefore, low pressure gases and medium pressure gases are two different supplies.

CORROSION

So far, this chapter has been concerned with the unplanned release of flammable gas or toxic gas or expensive gas. A further problem is that some special gases, for example hydrogen chloride, are extremely corrosive. This gas and others like it absorb moisture from the atmosphere. In the case of hydrogen chloride the resultant product is concentrated hydrochloric acid, which will attack valves and pipework. This creates an expensive maintenance problem even where the hazard to staff is contained.

THE UNIVERSAL SOLUTION

Neither of the two alternative schemes for supplying special gases is free from problems. There is no simple universal rule as to whether special gases should be kept outside the building with a permanent pipe distribution system, inside the building under the control of the user, or even whether they should be in the building whilst in use and stored outside the building when not. The decision has to be made on merit for each individual gas. The arguments for the various systems can be illustrated by considering two specific gases: nitrogen and hydrogen chloride.

Nitrogen

This is the most widely used of all special gases. In a chemistry laboratory it will possibly be required at every single bench and even other disciplines may require its availability in every room.

The number of cylinders would be very large if every user was supplied with a cylinder. This could block the freedom of movement of staff. In addition there is the possibility that in a fire each cylinder would over-pressurise and fail. Further there will be a large number of cylinder movements, each of which could result in damage or personal injury. Movement of cylinders may also result in damage to the valve with consequent projection of the cylinder akin to that of a rocket.

With a piped system there is the possibility of connecting a wrong cylinder to the system. In practice users are likely to opt for the use of liquid nitrogen for supplying the system thereby eliminating this hazard. On balance a piped supply of the liquid gas is favoured.

Hydrogen chloride

This is not likely to be a very widely used gas. Its use in the author's Company is limited to about two to three times each week in some part of the building. The demand in any given laboratory is, on average, less than once per month.

There are formidable technical problems in providing a piped system that will be free of corrosion. The system is apt to develop numerous leaks which can cause irritation of the skin, eyes, nose, throat and lungs of staff exposed to the gas, and be highly corrosive to laboratory equipment.

The problems of having cylinders of hydrogen chloride inside the building are much less than problems with nitrogen since the cylinders most commonly used for hydrogen chloride are smaller and lighter than those used for nitrogen. There is also likely to be fewer of them, probably only one or two per establishment. The balance clearly favours issuing cylinders to individual users as required and for these to be kept inside the building.

The removal of cylinders to an outside store each night could be considered but this entails problems. The reduction valve assemblies if left outdoors overnight could become cold and collect condensation when taken back into the building. This would result in the very rapid corrosion of reduction valves costing about £400 each. Consequently a cylinder could be taken outside only after the valve had been removed. This procedure would entail some 150 valve changes per year per cylinder. It is not unreasonable to assume that about once a year a mistake could occur causing cross-threading or valve damage which would result in leakage of this unpleasant gas.

It is the author's personal view that with this type of gas the cylinder should be kept in the building and moved as little as possible.

REFERENCES

[1] *Health and Safety Research 1980*, p. 15, Health and Safety Executive, HMSO, London, 1981.
[2] *The Fire on HMS Glasgow, 23rd September 1976*, Health and Safety Executive, HMSO, London, 1978.
[3] R. J. Friesen, *J. Chem. Educ.*, 1976, **53**, 373.
[4] *Fire Prot.*, 1979, **1**, 11.
[5] T. A. Kletz, *Health & Saf. at Work*, 1980, **2**(10), 42.
[6] *Piped Medical Gases, Medical Compressed Air and Medical Vacuum Installations*, Health Technical Memoranda No. 22, Department of Health and Social Security, HMSO, London, 1977.

27

The easy provision of telecommunications services in buildings

R. C. Rawling
British Telecom, Inland Customer Services Division

GENERAL

When designing or refurbishing a laboratory, most of the telecommunications requirements are easy to define. To achieve them will require considerable pre-planning by the architect, builder and telecommunications consultant.

The users' needs must be foremost in the plans. They require:

(i) Provision of all telecommunications services, i.e. telephones, telex, data transmission, either on a predetermined day or else very quickly. Completion before the predetermined day must not incur extra cost.
(ii) No disturbance to the furniture, fittings, decor or the work in progress in the building whilst the services are provided.
(iii) Flexibility in the type of terminal equipment to be used and the ability to change the layout within the premises, preferably without involving others, such as the telecommunications authority, to do the work.
(iv) Reliability supported by sensibly rapid servicing in the event of breakdown.
(v) Minimum cost both initially and as regular rental payments.
(vi) Conformity with the requirements of the Health and Safety at Work, etc. Act 1974 and other appropriate legislation.

Achieving these criteria poses many engineering and economic problems and the rapid change of technology must affect the decisions taken when designing new premises.

There are two ways of attempting to solve these problems.

(1) Traditional method
This uses a multiple twin wire cable to each terminating point radiating from a central position at which connection is made to a Building Distribution Frame

(BDF). Modern materials and methods promise significant cost reduction and the well-proven but old basic design has been re-born as an attractive competitor.

(2) Local area network
This is similar to a ring main but using co-axial or optical fibre, or occasionally twin wires, with tee-off at each terminal point. The highway carries communication information which is usually encoded digitally and each terminal is pre-programmed to extract its proper sections or packets. There are several commercial systems available and these are discussed later. The ring is not closed in some commercial systems.

CABLING OF BUILDINGS BY TRADITIONAL METHODS
Current United Kingdom Government policy has led to the liberalisation of British Telecom's previously held monopoly on a stage by stage basis starting with telephones and progressing to more complex devices, e.g. private branch exchanges (PBX). However, the implications for internal cable distribution have not yet been realised fully. A building's internal cabling which carries non-monopoly supplies only, i.e. PBX extension wiring, will soon be liberalised for competitive supply. Suppliers of such cabling, who will include British Telecom, will be free to tender for its planning and installation and to arrive at their own financial terms with customers, perhaps offering a choice between sale, lease and rental. Developers and architects should ensure that any such work is in accordance with the relevant Code of Practice currently being drawn up by the British Standards Institution[†]. This is because British Telecom will insist on compliance with this Standard before permitting connection of a PBX to its network. To enable British Telecom to compete for the work, each of its 61 Areas already employs a trained consultant engineer. To ensure financial fairness, British Telecom is expected to announce a reduced tariff for extension telephones that it provides via privately owned wiring.

Generous pair provision is a good economic investment. In a new building with no known occupier it is prudent to plan for one pair per 5 m^2 of floor space because this will meet all but the most unusual of building utilisation requirements. Planning of cables in the building requires close coordination between the consultant and architect so that riser ducts and trunking can be provided.

The connection of polyvinylchloride-coated wiring used to employ soldering and screw terminal techniques but British Telecom has now completely converted to insulation displacement hardware. The packing density at terminations has

---

†The code was issued in June 1984 and a certificate stating that the privately supplied wiring conforms to British Standard 6506 will minimise pre connection inspection by British Telecom.

also been improved about six-fold. The saving in labour costs is substantial and should be reflected in any quotation.

The creation of a circuit between floors is now made easily by cross-connecting pairs on the BDF. The connection of large numbers of circuits, for example, extensions from a PBX, is by the provision of an overlay link cable from PBX to BDF and thence to each floor. The connection of individual circuits is made by newly designed plugs and sockets which have gold-plated contacts and insulation displacement techniques terminations for quick connection. There is already a wide range available. Telephone circuits have been simplified. The old method used the transmission path in parallel, but the bell circuit in series. This has been changed completely to a fully parallel three-wire system making the use of many sockets on one line a practicable proposition. Traditionally these twisted pair cables were thought of as suitable only for analogue speech use; however, it has now been proved that they can support digital transmission at quite high speeds over considerable distances. The above adds substantially to the flexibility offered to laboratory designers and British Telecom has cooperated with many furniture manufacturers who already incorporate the new jack socket into their modular equipment.

The wiring of a building to give telecommunications outlet sockets is now easy and comparatively cheap but requires planning and coordination between user, architect and consultant. It does not in itself, however, provide complete flexibility under user control. However, when it is combined with a modern private automatic branch exchange (PABX), these ideals can be realised because there is then a user software control between the extension number and the port allocated.

LOCAL AREA NETWORKS

It is now quite practicable to exploit modern technology to provide a network within premises using a ring main technique. This requires some dedication and planning on the part of the user and the architect and the ability to make every user within the building compatible with the system. There is as yet no British Standard Specification on the method of utilisation of the network.

From the cabling aspect a single co-axial or optical fibre is looped through every terminal point and a tee-off established. The method of teeing varies between manufacturers but rarely does the cable need to be cut, and subsequent tee-offs are usually provided quite easily. The communication signals are generally encoded digitally, though some systems use an analogue multiplexed arrangement. Each and every terminal equipment must be pre-programmed to extract its own time slot or packet of information. Power to drive the terminal may be from a mains connection or over the co-axial pair.

There is no doubt that the future of such systems is exciting. They permit a terminal to be sited anywhere, yet it will still extract its own pre-programmed communication information. Some small overlay of traditional pairs for non-local area network circuits seems probable, for example, for the telephone in the lift, or the managing director's own exchange line.

The problems surrounding local area networks currently lie in the lack of recognised standards and the initial cost of provision; although the cabling is cheap the supporting hardware and software is often quite expensive. The user will be committed to the chosen system and may find this a constraint in the choice of terminal variety. The systems currently available are described in reviews produced by the computer industry. Potential users are recommended to study these.

The provision of communications services for laboratory or other purposes need not be an ogre. Many companies in Europe, Canada and the United States of America and national bodies such as British Telecom have specialists able to advise on the best way of meeting both short-term and long-term requirements. There is a range of circuits, for use both inside and outside laboratories, that have specified parameters and that can match any requirement for the transmission of telegraph, speech, music, data or video.

28

Meeting the evolving needs for computer and related services in an engineering laboratory

F. D. Riley and W. R. Hodgkins
Electricity Council Research Centre, Capenhurst

INTRODUCTION

This chapter describes how one particular engineering laboratory has set out to meet the evolving requirements for services in computing, data collection, transmission and analysis, and the various newer technologies primarily based upon or related to the computer. In the provision of any services it is necessary to take due account of the objectives of the laboratory and resulting requirements for the service in question. For basic services such as water or electrical power this may be a fairly straightforward task since there is considerable past experience on what is required and furthermore the technology in question is not changing rapidly over the lifetime of the laboratory. This is certainly not true of computer technology where laboratories set up some twenty or more years ago would make no provision for computers and today only the most reckless would predict with any certainty what form computer technology will take twenty years hence.

However, research itself is essentially long-term and therefore it may reasonably be expected that close analysis of the aims of the laboratory and of certain patterns of working will reveal targets and constraints which guide and assist in planning the provision of computer services. This indicates that there is unlikely to be a unique pattern which is optimum for all laboratories. Indeed no claim is made that there is one best pattern for each individual laboratory, rather that for an established laboratory there is a succession of choices which sets up a pattern which both opens up and restricts future choices. Thus it is of prime importance to see some overall plan, which itself is evolving, into which the individual decisions will fit.

RESEARCH OBJECTIVES

The Electricity Council Research Centre was established in 1965 to undertake research on new and improved methods of distribution and utilisation of electricity. About 350 staff are employed with a total annual budget in excess of £7 million. Of the staff, 90 are Research Officers, graduate scientists and engineers, and a further 120 are qualified technical support staff. The Centre has three main research objectives relating to three specific areas of activity.

(1) Industrial utilisation of electricity

This covers a very diverse area of activity including electrochemistry, electro-metallurgical processes, and electrophysical processes.

The general approach used is to examine various phenomena to see if these can be exploited to the benefit of electrical processes either already in existence or capable of being postulated. After a period of research and feasibility studies at relatively low levels of cost and research officer effort, a conclusion is drawn on the viability of proceeding to the development stage of the technology. This will depend on scientific and technical viability, likely costs of the process, and the likelihood of being able to find an interested collaborator or licensee for the development.

Once the development stage is reached, the project should be technically viable in principle, and the effort is devoted to building prototypes, small-scale pilot plants and mathematical modelling of the process. Having established viability both technically and economically at the pilot-plant stage, and probably also having established a patent, the emphasis moves towards collaboration with an industrial partner who is either already involved in that type of business or who wishes to move into that area.

After collaborative ventures have proved the principle of the process, a licensee is generally appointed to exploit the technology on behalf of Capenhurst.

At present (summer 1983) there are 127 current patents attributable to Capenhurst and, with the expertise amassed from the research effort, they form a valuable commercial asset which the Electricity Supply Industry intends to employ to the ultimate benefit of electricity users. Capenhurst at present has a significant number of licensees for industrial utilisation research projects in diverse areas.

(2) Environmental utilisation of electricity

This covers both domestic and commercial buildings and equipment.

The general approach has been to examine the total system in which people live and work, thereby directing research to fulfilling their fundamental require-ments of warmth and comfort at minimum cost through minimum utilisation of resources. As in industrial utilisation, a period of research is followed by exploitation of the techniques and systems developed through field trials on consumers' premises.

Exploitation of research in this area is related to a large market and, therefore, close liaison is maintained with the Electricity Council's Marketing Department.

Research in this area involves many disciplines, from pure chemistry and physics to environmental, psychological and social studies, followed by field trials prior to the exploitation of the systems.

(3) Distribution research

This covers the Area Boards' distribution system from 132 kV down to the consumer outlets.

The mode of research is rather different from utilisation. Research on distribution engineering is concerned with reducing the cost of providing the service while maintaining the quality and continuity of supply. The work strikes a balance between solving long-standing problems, such as improved gale resistance of overhead lines, to new developments in microprocessing and microcomputers. The emphasis is very much on exploiting the power of the 'latest' tools for solving problems.

Much of the research results in new or modified equipment which is then licensed to manufacturers who may sell the equipment to Area Boards but who also sell abroad. This is just another aspect of Capenhurst trying to help to keep United Kingdom industry up-to-date and competitive in world markets.

A new area of activity is contract research which Capenhurst considers to be a profitable way to improve the interaction of research staff with United Kingdom industry and commerce. External contracts for research are undertaken which are mutually beneficial to United Kingdom industry and to the Centre. To date, this has been a very successful venture and many research contracts have been undertaken, are under way, or are being negotiated.

IMPLICATIONS

Constraints

It will be evident from this description of the objectives that several factors, some of them peculiar to the Centre, must be taken into account. The first is that there is no one major application. In fact there are more than 100 separate research jobs, with a few of the larger projects employing up to about six technical staff. Furthermore the work covers an amazingly wide range of activities and disciplines. These factors imply that for most projects there is unlikely to be a member of the team, if indeed it is a team, who is very knowledgeable about the latest developments in computing. Neither is it possible to assign a specialist in computing to each project, especially one with any depth of knowledge of the particular field of application. Another factor is the great importance attached to the eventual ability to transfer the technology to outside industry. Thus projects must not become dependent on skills or facilities which only pertain to the Centre. Indeed a major part of the project work of the Centre, which includes much of the collection of experimental data, is performed outside the Centre and often in situations where no other laboratory facilities are accessible. How this works out will be illustrated later.

Another factor of some significance not apparent from the objectives is that the Centre consists of several separate buildings, some dating back about twenty years prior to the setting up of the Centre and others of more recent, purpose-built construction. These buildings are scattered widely over the site. This at least has the advantage that planning to take account of off-site work is not so very different in concept from dealing with some of the on-site work.

Research work is not of course only dependent upon objectives and facilities. The most important input is the staff. When a laboratory is set up, and during its expansion, it is relatively easy, initially, to plan facilities and then recruit staff to man them. Even so to recruit and maintain good staff in the rapidly expanding field of computers may not be a trivial task. In contrast the Capenhurst Laboratory has been fully staffed for almost 15 years and few staff are near retirement. Furthermore the Electricity Supply Industry, in common with many other basic industries today, is aiming, as efficiency improves (often as a consequence of the research undertaken), to shed staff rather than recruit them. Thus it becomes virtually impossible to change the balance of skills within the laboratory by recruitment of new staff. Therefore, it is vitally important to ensure that the facilities provided are suitable for use with the existing staff so far as they can be retrained in their use. In particular it is found that there is a lack of trained staff in computing and electronics, and even more so of those versed in both. This is a most difficult gap to bridge. Only a few of those skilled in software are prepared to, or perhaps have the aptitude to, work closely with hardware. Those trained in electronics are very often reluctant to appreciate and use modern software techniques.

Two final constraints should be mentioned here, although neither has been predominant in governing the path taken at Capenhurst. The Research Centre is part of the wider Supply Industry and it too has its plans for similar and occasionally overlapping facilities. The Centre services a large number of virtually autonomous customers, the Area Boards, and the Council itself has not had centralised computer facilities, consequently no real restraint has been suffered in this respect and planning has never been significantly delayed impending some larger plan. But, as will be seen, an attempt has been made to take account of the needs of other parts of industry. In future it is likely that computer facilities, and more particularly communications, will be linked more closely with the Council headquarters, but the idea that there can be one great plan encompassing every detail should be treated with the utmost caution. Rather, the plan should be flexible and able to accommodate a variety of local implementations.

The second factor is of course cost. Every organisation has the aim of providing the best facilities at the lowest cost, and the cost of the various choices has always to be taken into account. However, in practice none of the Centre's major decisions has presented any great difficulties in justifying on cost grounds and, overall, the proportion of the budget spent in data processing is small, especially when compared with its importance and heavy use in many of the research projects.

Although many of the constraints are of course peculiar to a particular laboratory, they are likely to resemble those faced by many other established engineering and research laboratories. If they seem somewhat different from those faced by a new or expanding laboratory then it must still be remembered that it will not be long before one is living with one's own history.

Aims

A more detailed look at the aims in providing services in data processing and related areas is now necessary. Needless to say the aim must be to provide first-class facilities in the sense that what is provided must meet the real needs of the research and of the staff as fully and as flexibly as the state of the art will allow. Naturally this is qualified by all the restraints described above, but the key word here is 'flexibly'. The service must be responsive to user needs and also anticipatory. Many users will not realise the potential of technology without being led into it. Given the many people who have little familiarity with computer technology, it is vital, in order to gain widespread use, that the facilities be user-friendly whilst at the same time providing a first-class service for those who are professionally more familiar. Users must feel free to make their own mistakes without fearing instant criticism and systems that seek to monitor and guide the users too closely have the effect of deterring users from ever learning how to carry out anything for themselves.

Another clearly stated aim is to minimise the use of specialist staff. Besides the constraint that they are not likely to be available this has the positive advantage of seeking to make the users as responsible as possible for their own work and thus most likely to identify their real requirements. Furthermore it minimises the division between them, the users, and the specialists who provide the service. It also avoids the setting up of an extensive centralised computer department incurring not only large overheads but the danger of becoming self-perpetuating with its own objectives, gradually diverging from those proper to the Centre. It is never easy to recover from this once it has happened and it often ends up with users setting up their own separate systems with much needless duplication of both hardware, software and human resources.

It may be questioned whether the aforesaid aim of minimising the use of specialist staff may not conflict with the general principle in conducting research, which is to free creative research staff from peripheral activities in order to concentrate on the science of their projects. Thus it is important to ensure that the tools provided by the computer are simple and direct to use and do not involve unnecessarily detailed expertise as is sometimes required in procedures for file-handling or job-control languages.

Another of the key aims is to avoid needless duplication of effort. Although some work is similar, inevitably much is different. It is therefore important to choose the most appropriate level to provide common systems which do not unduly restrict the user and yet are of wide use. One example is described later for providing a degree of automatic analysis of experimental data. Others are in choosing standard interfaces at both hardware and software levels.

Finally it must be remembered that no laboratory is or should be self-contained. Staff must be encouraged to know about and be familiar with external resources which it is uneconomic for the particular Centre to supply internally. Thus decisions have to be made, for example, on what programmes or packages should be transferred to an in-house computer and which should only be available from a bureau. Planning for this, to provide good enough access to bureaux and their expertise, is not always as simple as it seems.

HISTORICAL DEVELOPMENT

For several years Capenhurst operated in much the same way as any industrial research laboratory, doing background research and bench scale work and building rigs in a fairly standard and sequential manner, with detailed analyses of the results between each stage. The computing aspects were mainly dealt with by a team of a few specialists, most of them mathematicians, who used a remote job entry terminal to a large mainframe at a computer bureau. The bureau was able to provide an efficient service with a quick turnround and, of course, if the service deteriorated there was a real option of placing the business elsewhere. Some dedicated minicomputers for handling experimental data in conjunction with data loggers were also provided.

By 1978 expenditure at the bureaux was increasing to the point where it appeared worthwhile checking the economics of providing in-house facilities. At the same time there was an increasing demand on site for desk and mini-computers to provide local or interactive features. It was clear that unless some decision was made on an overall plan there would be a large-scale introduction of possibly incompatible local computer systems which it would be impossible to support centrally and which would grow, at considerable expense in man effort and equipment cost, so that each would ultimately provide quite powerful, expensive and duplicated facilities.

The initial proposal therefore was to provide a central interactive computer which would be able to carry out about half the existing batch work with the remaining half of large computing problems still being carried out at the bureaux. Terminals would be provided in each building, as required, linked at a minimum speed of 2400 baud by dedicated telephone lines. Links between buildings would be rented initially from British Telecom but the opportunity would be taken to lay the Centre's own cables as opportunity presented itself. The aim was to be able to support at least 15 interactive users simultaneously enabling about 25 terminals and devices to be attached to the system allowing for the fact that all terminals would not be used simultaneously. In line with the philosophy described under 'Implications' it was seen as very important that the system be open to all potential users and be suitable for computer novices as well as providing at least as good a service as was available at the bureaux to the existing experienced users. The system would therefore provide BASIC for simple use, a structured FORTRAN, in fact FORTRAN 77, for larger scientific users, plus good editing and file handling facilities for general work. In addition it would need a batch stream to handle existing batch programmes, and emulators to enable links to be established into any of the main computer bureaux. It was anticipated that with such a system the demand would grow rapidly and any proposed system must be readily capable of expansion.

These proposals were accepted by management for inclusion in the 1979 budget. After a thorough market evaluation of the many alternatives, a Prime 650 computer with 0.5 Mbyte of semiconductor memory, 16 interactive ports and two 80 Mbyte disc drives was selected and came into operation in September 1979. Rapid developments in technology since the proposal had been put forward meant that the computer was more powerful than that which it had originally been possible to afford and, within a short time, about 90 per cent of the bureau work was being run internally. However, the forecasts that the number of users and amount of work would grow was rapidly confirmed and the system has since been expanded to support up to 32 interactive lines, although this would seriously degrade the performance of the computer if all were heavily used, to a memory capacity of 1.5 Mbyte, and a 300 Mbyte disc has been added. Currently there are well over 100 registered users and during a single week the system is typically used by over 50 of them. The computer usage has increased by about six times over about three years, reflecting not only an increase in use by the previous users but the much more widespread use on an enormous variety of work by the many new users.

One particular area of growth has been in the analysis of data from experimental rigs, and chemical and physical analysis equipment. In the past most analysis equipment used a stand-alone machine which displayed numbers on dials or scales. Later printers were linked to them, but now this type of equipment is specified with a compatible interface to the Prime 650, either directly or through an associated cassette reader or microcomputer, so that the results can be analysed and presented, usually in graphical form.

Another large area of growth has been in graphical presentation of results. For this purpose there are a variety of plotters and terminals with graphics features available and more importantly some common software written for analysis and plotting so that any of several different devices can be used.

IMPACT ON METHOD OF WORKING

Having described some of the technical aspects of the computer revolution, it is well worth considering what impact this has made on the way research is conducted.

In the old days (five years ago!), research was very much conducted at the laboratory bench, after reviewing the literature or following up bright ideas. When results were obtained on a small scale, the Research Officer would analyse and interpret them and decide how to move forward or, if the results were not too good, backwards. The difference that interactive computing has made is really quite remarkable. The Research Officer still needs his bright idea to work on but having obtained a few results and with some understanding of the physical or chemical phenomena being studied, it is often possible to construct a mathematical model which produces as many results as required. These are then checked at various levels of statistical significance, and a design produced for a rig. This can be checked against the model and optimised for various key parameters. Also, if the rig requires modification then questions can be asked of the model to see what effect may result. This saves a considerable amount of time in optimising rigs. When proven, this same mathematical model can be used as the design guide for specifying industrial plant.

Finally, because most of the data is either stored on the computer or is readily accessible to it, it is now much easier to present results in tabular or graphical form. These can be incorporated in reports without the considerable amount of manual typing and drawing previously required.

EXAMPLES OF PROJECTS IN WHICH THE COMPUTING FACILITIES HAVE BEEN USED

In order to illustrate the type of work that is carried out at Capenhurst and to show how this is affected by the availability of interactive computing, three examples are discussed.

Heat pumps

A recent project concerned the development of a heat pump dehumidifier to operate as a timber dryer at temperatures up to 80°C, well above those previously available. The project team consisted of several engineers and physicists together with a mathematician and a chemist. A key part of the project was a mathematical model which was used to optimise the design of the heat exchangers and airflows through the machine. For this the computer was invaluable as it enabled a large number of possible configurations to be considered before the best was chosen.

The dehumidifier itself could not be considered in isolation, but had to be analysed as a component in a total timber-drying system. Thus an approximate model of the heat flow and moisture flow in different timbers was required from which an integrated model of the total energy flow in the system could be built up. This enabled estimates of the running cost of the kiln to be established thus forming the basis of an economic case for prospective purchasers.

After the building of an industrial-scale demonstration kiln, considerable use was made of data logging and processing facilities. These enabled a detailed analysis to be made of the heat pump dehumidifier itself and of the overall system, and provided information to feed back into and confirm the validity of the computer models. The data processing was accomplished by the Capenhurst general-purpose data analysis software – the function of which is to provide a quick, compact graphical display of large quantities of experimental data, having first checked and, where possible, removed errors arising in the collection and recording systems. Coupled to this general-purpose software was the computation package, necessarily different for each application, which in this case made use of such information as the thermodynamic properties of the refrigerant fluid and the psychrometric properties of moist air.

At the end of the project the energy consumption model was condensed in order to provide a simpler but still reasonably accurate model which could be easily run by the sales engineers of the firm to whom the technology was transferred.

Low energy house

The aim of this project was to prove the practical and commercial viability of electrically heated, low energy houses. This stage represents the culmination of an extensive research programme at the Centre on energy flows and heating requirements of very well insulated houses.

A deliberate decision was made to stretch conventional building techniques rather than introduce unfamiliar experimental methods of construction. Two national builders agreed to build and sell two houses each, chosen from their standard range but modified to incorporate special features such as improved thermal insulation and controlled ventilation.

An essential feature of this exercise is to monitor the thermal performance of the houses both before and during occupation. Predicted and actual energy consumption have to be compared regularly so that discrepancies can be investigated immediately.

Sensors were installed during construction to monitor air temperature in each room and the setting of the ventilation system. The data logger records the readings half-hourly on a magnetic cassette tape. A second data logger records itemised electricity consumption and the opening of doors and windows. A weather station on the housing site provides weather data on a third logger.

Data collection has to be carried out with minimal disturbance to the occupants. Hence the cassette recording system chosen needs no daily monitoring but only the changing of tapes once each week. The tapes are returned to the Centre for data processing and graphical display using the same suite of programs described in the previous example. This allows a rapid visual assessment of the week's trends. Afterwards a calculation of predicted and actual energy usage is available from a program which was developed separately. It is regarded as essential that any failure of equipment or disturbing trends can be recognised within the week in time for the following visit. Allowing for the travelling time this allows only three working days to read in the ten or so tapes and carry out the analysis. Indeed the volume of data is so great that any failure to complete the analysis on time causes problems in storage in readily accessible form on the computer.

Because of the provision of the standard interfaces for the loggers and the graphical display program, this project required very little computer specialist time, the programming for the energy analysis being carried out by the end user.

Transverse flux

Transverse flux induction heating appears to be a very attractive method for annealing and other heat treatment of metal sheet products. Although first proposed in the 1930s, and the subject of several development projects since, transverse flux induction heating techniques have never been commercially exploited in the past because of the difficulties in obtaining a sufficiently uniform temperature distribution in the heated metal sheet.

Recognising that there were fundamental problems in devising a suitably contoured magnetic field distribution in the heater, a major project was started in 1972 to investigate these magnetic field problems and produce a commercially viable heating unit. An electrical engineer, well-versed in the analytical study of electromagnetic field problems, and with extensive experience of computer techniques, worked full time on this project to obtain sufficient understanding to design an appropriate hardware system to control the magnetic flux. This involved writing a series of programs to solve particular electromagnetic and thermal diffusion problems, making the most of physical features of the heater to overcome the difficulties in solving true three-dimensional eddy current equations.

During the experimental stage of building the hardware, problems were encountered in logging and analysing the data produced during the very short tests of about five seconds duration that were necessary to develop the hardware fully. The brevity of these tests was due to the use of short lengths of metal pulled through the heater to simulate continuous heating of coils of metal strip. This required careful programming of loggers to operate at maximum speed, and the careful writing of data analysis programs to obtain useful experimental data in tabular or graphical form as required. These programs were complicated by the necessity to dispense with all unnecessary characters on the paper tapes used as a data storage medium for loggers of that period.

The next stage of the project involved the building of a full-scale development rig at a metal processor's works. Extensive data logging facilities were provided on the rig and as development proceeded paper tapes were superseded by magnetic cassettes as a data storage and transference medium. These tapes were brought back to the Research Centre for processing. An additional feature of the development rig was the use of software programmed sequence controllers. This required the learning of the appropriate programming languages, and the on-line development of these programs as new features were added to the plant.

The final stage of the project, now in progress, is the transfer of the technology developed over the ten-year life of the project to licensees, who will design, build and market the fully developed heater. The licensees will need to run versions of the original electromagnetic field solution programs to ensure that the designs are accurately produced. These heater installations are typically rated at one to five megawatts; they cost several million pounds, and no deficiencies in the design of these novel devices can be tolerated. The computer programs to be handed to the licensee can only be run on the largest of computers and must therefore be installed at a major computing bureau. Experience of the interfaces with such bureaux is therefore necessary, both to test the programs and to train the licensee in the use of the programs, accessed from their own design offices.

It is apparent from the experience gained from this project, which has lasted some ten years, and seen several developments in computer and associated technology, that attempts must be made at all times to devise sufficiently general software so that the transfer to the latest computing and data storage devices can be effected with the minimum waste of time and effort. When it is remembered that it can take many months of use to make a program fool- and bug-proof, changes to new computing hardware systems must not be allowed to reintroduce these gremlins. Another factor has been the need to provide adequate development facilities at the Centre to allow off-site trials to proceed at reasonable speed. Here the ability to interface different computer and data systems is becoming increasingly important.

STAFF

It has already been stated that the choice and success of a policy must be affected by the staff. The present system at Capenhurst runs with one full-time systems programmer and one full-time operator, an increase of half a post from that required to run the previous batch terminal. Another junior programmer acts as reserve operator and some guidance on hardware is available from a Research Officer. Thus the aim of minimising the service staff has been met. The more-experienced users are able to provide some guidance and assistance to the less-experienced users both on an individual level and by providing short introductory courses on various aspects of computing. On the whole, however, those who have the interest and willingness to try it have had few problems and have welcomed the system.

APPLICATION TO SUPPORT FUNCTIONS

The Prime 650 computer was purchased as a scientific tool to be used mainly in an interactive mode by research and technical staff. No direct provision was made for commercial work or management information systems, but because of the open computer policy several applications have been or are being developed.

A start has been made on a computer-aided draughting system to enable drawings to be stored digitally by installing a proprietary software package. A database of *Capenhurst Research Reports* has been made accessible to Headquarters and to Area Boards using the Prime package 'POWER PLUS', and consideration has been given to utilising the Prime 650 system for providing management with accounting information.

Other facilities are provided by a word processor linked to an Optical Character Recognition machine for production of reports, a Viewdata intelligent terminal, and terminals for access to worldwide databases including Europe and the United States of America through IPSS, Euronet and PSS networks.

COMMUNICATIONS

As mentioned earlier, the Prime 650 is linked to a large number of terminals on site. Asynchronous access may also be made through either the public switched network or the Electricity Supply Industry's private network, thus enabling Area Boards or staff working off-site to work directly with programs and data on the computer. Interfaces and software have been developed to allow the transfer of files readily from various other minicomputers on site, including the DEC LSI11 range and APPLE computers.

In 1983 the first phase of a communications network was introduced. This allows data to be transmitted over existing telephone lines to a data switching unit concurrently with voice traffic. Eventually this will enable data to be sent from any telephone extension to the computer and allow any devices conforming to a simple protocol to be connected to each other.

Electronic mail links between word processors at Headquarters and local word processors, and the provision of electronic typewriters capable of direct input to the word processor are under development. The linking of the Viewdata intelligent terminal to an in-house video system for preparation of graphics artwork and direct input to the video recorder through a video editing system is also being developed.

THE FUTURE

Looking a few years ahead it is clear that many of the things for which users of the new technology have and are constantly striving for will become standard features. Digital communication is now starting to look healthier. Although not quite the case at the moment, it appears that both British Telecom and private suppliers appreciate the need for interfaces and standards that are clear, and available. The power and versatility of the microcomputers and intelligent terminals will continue to increase and this is especially welcome if the ability to interface and interconnect such devices is attained. Computer languages are becoming standard and provide useful development features and this is likely to continue to improve.

In the area of office technology there is rather more confusion, but here if standard interfaces to a digital PABX become common, many of the current problems may be alleviated.

ACKNOWLEDGEMENTS

The authors are indebted to many of their colleagues for their assistance in the preparation of this chapter. In particular mention should be made of Mr F. W. Sharman, Dr D. J. Dickson, Dr R. C. Gibson and Mr D. Cross.

29

The handling and storage of flammable and toxic materials – a university safety officer's viewpoint

G. Hargreaves
Imperial College of Science and Technology, London

INTRODUCTION

A university environment presents special problems because of the highly varied nature of the work undertaken by an academic community. Hazardous work is often located in the middle of other activities. Each activity in itself may not be particularly dangerous, but in combination, can lead to a chain reaction of events, and a possible disaster. In the main, the free exchange of academic information leads to a parallel exchange of safety information although the work is sometimes undertaken without due weight being placed on the safety aspects; the excitement generated by discovery outweighing all other considerations. Too often a research programme is presented for funding without the safety element being considered and budgeted for. Many outside bodies will not consider funding for safety aspects of proposals. In the author's opinion pressure should be put on such bodies to bear in mind safety needs when awarding grants, especially in the light of present legislation and current financial restrictions.

Research activities start and then fall off as the work develops. This changing and alteration of course is an essential part of most research. It does, however, lead to short cuts and the jettisoning of apparatus, materials, etc. Consequently, a system has to be devised whereby departments within a college or university are self-monitoring, the system being designed to identify problem areas and to recommend action on these matters to the executive bodies.

Pressure can be put on the individual by using a system of committees which provide the expertise necessary to monitor and, if need be, advise on safety aspects of work.

Figure 29.1 is a diagrammatic representation of typical arrangments made by universities. Some of the relationships are advisory, some executive. Departmental committees control activities to a large extent by means of safety audits,

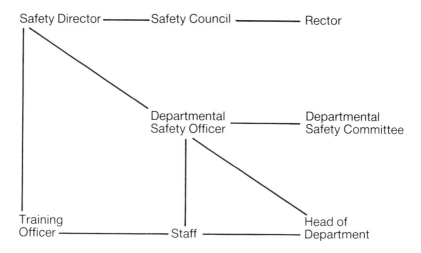

Fig. 29.1 – Typical university safety organisation.

carried out on a regular basis. The inter-relationships shown are aimed at stimulating such inspections and results are monitored by the Departmental Safety Committee as well as the full College Safety Council via the Safety Director. Accident and Dangerous Occurrence Reports are also examined and investigated.

EXTENT OF THE PROBLEM

Flammable and toxic materials accumulate for many reasons. Over-ordering is common because of the autonomy of many groups or departments. At the end of projects or when staff or students leave, apparatus and associated materials are often abandoned. There is a reluctance to use 'old' materials, often for very good scientific reasons; but too frequently chemicals are presented for disposal when they are in perfectly good condition.

HANDLING AND DISPOSAL OF WASTE

Central organisation of waste disposal is one part of the solution to the problem, but this can only be effective in large communities with varying activities if the committee structure, shown in Fig. 29.1, works satisfactorily. From the pressure exerted by committees, a Safety Officer can have, in effect, an executive function despite the frequently asserted statement that Safety Officers are purely advisory. It is a question, very often, of persuasion and organisation. The latter not only includes practicalities such as making disposal arrangements and carrying out the paperwork, but also training people, in the broadest sense, to react to the arrangements.

LABORATORY DESIGN

The present laboratory design standards for scientific laboratories leave a great deal to be desired if the academic community is to have the freedom to work flexibly. No amount of pressure from committees, departmental safety organisations and individuals will stop workers from accumulating and using flammable and toxic materials in their own way. The concept whereby large volumes of materials are kept in stores outside buildings and brought into laboratories in small quantities, does not seem to fit in with the way in which human beings can or are prepared to work. It is inefficient and needs ancillary and technical staff and this is not possible in the present financial climate. It would require a police type of monitoring system to prevent workers from stocking up with large quantities of flammables, etc. A series of small fire-proof stores built as part of each laboratory would answer most needs. By restricting the size of such stores to that acceptable to the fire authorities, a solution satisfactory to both parties would be achieved. Only the discipline of returning flammables to stores when not in use and overnight would then be necessary.

In general, coping with the many problems would be easier if services and other facilities were more readily altered and more accessible. If a building such as the Pompidou Arts Centre in Paris can be designed with all its services on the outside and its activities on the inside, why should universities and technological institutions be built with inaccessible services inside buildings, tightly sandwiched between corridors and laboratory walls? The occupier would be able to meet the requirements of the Health and Safety Executive and other bodies to store gases outside the building and pipe them inside if an 'outer skin' of service ducts was used.

At the present time there is increasing difficulty in disposing of the small quantities of waste flammable solvents, chlorinated solvents, and other toxic materials because new legislation has led to excessive bureaucracy. The problem now experienced is that disposal companies do not find it worth their while to dispose of small quantities of widely varying waste because of the cost factor. Is it beyond the imagination of architects to build in systems for dealing with waste? Animal incinerators are common in biology buildings because of the Home Office laws. It seems not unreasonable, therefore, that every chemistry department should have some form of chemical incinerator provided with the adequate scrubbing and other facilities necessary to reduce pollution.

SUMMARY

The problems of universities are increasing, not necessarily as a result of their own activities but because of the pressure from outside. There is a shortage of money; but even if there were plenty, there would still be difficulty because of the restrictions and controls which make disposal more and more problematic.

The only solution is for universities to have the facilities to cope with their own waste locally. In considering university designs of the future, architects and academics must give greater weight to the problems of waste disposal, and this requires imaginative design and local installations capable of dealing with relatively small but varied quantities of waste materials.

30

The disposal of flammable and toxic wastes

R. C. Slade
Queen Elizabeth College, London

INTRODUCTION

All laboratories generate waste products. Some of this waste will be hazardous that is, material which is highly flammable or toxic and which is now classed as special waste under the Control of Pollution Act 1974. Special waste creates problems for laboratory designers and architects as well as for laboratory managers, supervisors and workers.

The difficulties associated with special waste are not necessarily those of scale but also of complexity. It is frequently easier to deal with large amounts of waste if its composition is known and unchanging than with small amounts of unknown materials or materials of varying composition. The scale and nature of the waste produced by a laboratory depends upon three basic factors – the size of the laboratory (ranging from a small analytical section in a manufacturing plant to a large multidiscipline complex), the type of laboratory (school, university, medical, industrial) and the activities of the laboratory (teaching, research, quality control). Any laboratory is likely to be faced with waste products containing biological, chemical, clinical, or radioactive materials in the form of solids, liquids or gases. The hazards may be those of flammability or toxicity; if the latter, acute, chronic or infectious hazards may be present.

DISPOSAL METHODS

There are several well-established methods of disposing of waste materials, each of which is regulated by Acts of Parliament, regulations or local bye-laws. These methods are:

(1) Into the sewers
(2) Into the atmosphere
(3) By licensed contractor
(4) On site

Disposal into the sewers

Laboratory waste discharged into the public sewers is classed as trade effluent and as such must be carefully controlled within the restrictions imposed by the Water Authorities. With some exceptions these restrictions are contained in regulations made under the Public Health (Drainage of Trade Premises) Act 1937. Generally, disposal is restricted to water soluble materials provided that the concentrations of certain substances conform to local regulations. Flammable and highly toxic substances are forbidden, although in practice small quantities of water-miscible liquids, e.g. acetone and ethanol, may be permitted. Some substances may undergo reaction in the sewers or drains to produce hazardous products and these also must be excluded; examples include azides which may react with metal pipes and cyanides or sulphides which react with acidic effluent.

Disposal into the atmosphere

It is tempting to dispose of gases or vapours by releasing them untreated to the atmosphere. Where small quantities are disposed of on an infrequent basis this may be an acceptable method for flammable liquids and toxic volatile solids. However, if the regular release of gases is necessary then the Clean Air Acts 1956 and 1968 may impose certain restrictions enforced by the Local Authorities. Thus scrubbers may be needed to reduce the acidity of gaseous discharges. In a residential area the release of toxic or obnoxious substances will almost certainly lead to action by the local inhabitants.

Disposal by licensed contractor

Unless special facilities are available, untreated flammable or toxic wastes will often be disposed of by this method. The Control of Pollution (Special Waste) Regulations 1980 impose stringent requirements on producers, carriers and disposers, and laboratory managers will need to set up an effective organisation to deal with these. Facilities may be needed to analyse mixtures of wastes so that adequate descriptions can be provided to the disposer. The waste must be stored safely at the laboratory prior to collection, and there must be an administrative back-up to deal with consignment notes and to keep proper records.

Disposal on site

Some laboratories by virtue of their size or activity may find it convenient to make provision on site for the disposal of special waste or for its pretreatment prior to final disposal. Thus many hospital laboratories will have access to an incinerator for the disposal of waste containing infectious material; other laboratories may treat wastes to recover expensive components which can be re-used. The role of the laboratory designer, in consultation with the user, is to advise on and provide any special facilities which are needed for on site treatment. In providing this advice the designer must be aware of statutory requirements in related buildings and activities. Many laboratories are excluded from the requirements of the Factories Act but the Act and its associated regulations provide a useful code of practice. Thus the Highly Flammable Liquids and Liquefied Petroleum Gases Regulations 1972 require that highly flammable liquids be disposed of either in plant or apparatus suitable for the safe burning of that liquid, or be burnt by a competent person in a safe manner and in a safe place. This requirement should be applied in all laboratories where facilities for the disposal of flammable liquids are required.

METHODS AVAILABLE FOR DISPOSAL ON SITE

Autoclaves

Any laboratory handling infectious or biohazard material should have direct access to an autoclave for the sterilisation of waste prior to final disposal. For effective sterilisation the autoclave must be designed and operated such that the load temperature is monitored by carefully positioned thermocouples; the Medical Research Council recommends[1] that the load be maintained at not less than 121°C for not less than 15 minutes.

Macerators

Some biological waste, including animal carcases, can be disposed of into the public sewer after maceration. Carcases containing low level radioisotope tracers can be disposed of in this way but infectious waste must first be autoclaved. Macerators need to be installed where there is an adequate water supply and where connection to the drain avoids sharp bends or long horizontal runs.

Incinerators

Incinerators are commercially available for the disposal of a wide variety of waste materials ranging from animal carcases to flammable solvents and household rubbish. The design of the incinerator will depend on the type of waste to be burnt and suppliers should be consulted on the particular requirements of the laboratory. A general-purpose incinerator suitable for laboratory waste should incorporate the following basic components.

— A *combustion chamber* lined with refractory brick into which solid materials can be introduced either manually or automatically.
— An *atomiser* for feeding liquid waste into the combustion chamber, together with a holding tank and pump or gravity feed delivery system.
— A *secondary or after burner* to ensure complete combustion of carbon or other deposits.
— A *high velocity discharge stack* to release cooled gases clear of any surrounding buildings.

The complete destruction of flammable and toxic wastes by on site incineration is attractive for several reasons. There is economy of operation, particularly if heat recovery is available, compared with disposal by licensed contractors. The accumulation of hazardous wastes on the premises is prevented and storage problems reduced. Much of the administration and paper work associated with the use of contractors is removed.

There are, however, some disadvantages and limitations that should be considered at an early stage in the design of the laboratory facilities:

(i) Inorganic compounds are converted to oxides which are discharged from the stack or accumulate in the ash; in many cases these metal oxides are highly toxic.

(ii) The temperature in the combustion chambers must ensure complete oxidation; for many thermally stable compounds temperatures of 800 to 1200°C are necessary.

(iii) Many waste materials are non-flammable or have low calorific values; in these cases it may be necessary to mix wastes to ensure even combustion and total oxidation of all the components of the load.

(iv) The decomposition products of halogenated chemicals are highly corrosive and extremely toxic; products such as hydrogen chloride may need to be removed by passing the discharge through extensive scrubbing or wash-down facilities to prevent the rapid corrosion of the incinerator.

(v) Although certain low level radioactive wastes can be incinerated with authorisation under the Radioactive Substances Act 1960 there is a possibility that the furnace lining and discharge stack may become contaminated.

(vi) Where plastics waste is burnt in a general-purpose incinerator the amount should not normally exceed 15 to 20 per cent of the total load.

(vii) The incinerator should be located well away from other laboratory buildings or facilities. If this is not possible then adequate fire compartmentation is necessary.

STORAGE FACILITIES

Irrespective of the methods which a particular laboratory employs to dispose of its special waste some accommodation will be needed to store the waste until it is removed or destroyed. In the case of flammable liquid waste, including petroleum spirit or mixtures thereof, a separate part of the main solvent store may be set aside for this purpose. The store itself will have been approved by the Local Authority in the case of a licensed petroleum spirit store, or by the Factory Inspectorate where highly flammable liquid is stored on premises subject to the Factories Act. If a separate store is to be constructed to hold flammable waste then the relevant regulations will apply as they do for general solvent stores and the appropriate authorities should be consulted when plans are being drawn up.

Toxic wastes may need to be kept in stores which are both safe and secure. Ventilation, e.g. a ventilated store cupboard or fume cupboard, may be a requirement of such a store which should be located away from the main activity or process areas. As with any other chemical store, equipment should be provided to deal with any spillage.

USER REQUIREMENTS

No system for laboratory waste disposal will operate efficiently unless the staff and students or pupils understand the system and are provided with the necessary equipment and facilities. In chemical laboratories containers will be needed to allow different wastes to be kept segregated from one another, e.g. halogenated and non-halogenated solvents, and to collect specific residues from different processes. The segregation of laboratory waste is important not only to avoid the mixing of incompatible substances but also to facilitate any subsequent treatment or disposal. Containers for flammable liquid waste should be chosen with care because of the obvious fire risk if vapours are allowed to escape freely from the containers; cans fitted with flame arrestors and spring-loaded or fusible-link hinged caps will significantly reduce this risk. If metal containers are chosen it will be necessary to consider the corrosive nature of the liquid waste.

In biological or medical laboratories sealable plastic bags should be available for infected and other biohazard wastes and these should be removed for autoclaving on a daily basis. The code of practice for clinical waste[2] makes several important recommendations including colour-coded bags for particular items. Materials in contact with toxic chemicals or infectious substances — swabs, disposable laboratory ware, tissues, hypodermic syringes and needles, broken glassware, etc. — should always be segregated from the domestic waste and containers for such items should be available.

Any container which is provided as a receptacle for flammable, toxic or other laboratory waste must be clearly and adequately labelled to describe the nature of the waste and its principal hazard. The transport on the public highway of special wastes from producer to disposer may invariably require some form of labelling on the containers or vehicle; if the laboratory uses its own transport then adequate insurance cover must be taken out.

ORGANISATIONAL REQUIREMENTS

The provision of appropriate facilities and arrangements for dealing with special waste is a responsibility that falls on laboratory managers or supervisors. Adequate procedures for disposal must be set up and brought to the attention of all laboratory workers. These procedures should cover the following points:

(a) Allocation of responsibility within each laboratory or department.

(b) Nomination of a competent person to administer the organisation.

(c) Segregation of various types of waste.

(d) Clearly labelled waste receptacles.

(e) Treatment of highly toxic wastes before disposal.

(f) Removal of waste from laboratories.

(g) Storage of waste prior to collection or disposal.

(h) Arrangements for on site disposal.

(j) Arrangements for liaison with licenced waste disposal contractors.

(k) Information to all personnel on all the procedures and arrangements which form part of the waste disposal system.

Some laboratory personnel see waste disposal as a non-productive activity which interferes with their research or other essential work. However, the collection and disposal of special waste is an essential part of efficient laboratory organisation and management. Facilities for the disposal or treatment of such waste are as important as those provided for all other laboratory activities. The pressures on laboratories to act responsibly in dealing with their waste are increasing whether from central government, local authorities or environmentalists. Laboratory designers, managers and safety officers need to work closely together if these responsibilities are to be met.

REFERENCES

[1] *Annual Report,* Medical Research Council, London, 1959.

[2] *The Safe Disposal of Clinical Waste,* Health and Safety Commission, HMSO, London, 1982.

Editors' Note

A new book entitled *Handbook of Laboratory Waste Disposal* by Martin Pitt is to be published by Ellis Horwood Limited in 1985.

Section 3:

SPECIALIST LABORATORIES

31

Microbiological safety in laboratories

H. B. Ellwood
Building design Partnership

INTRODUCTION

Pathology, the study of disease, embraces four main disciplines:

(1) Morbid anatomy – the study by dissection of the diseased human or animal body.
(2) Haematology – the study of blood, its composition, structure, functions and disorders.
(3) Chemical pathology – the study of the chemistry of the living tissues and fluids of the body.
(4) Microbiology – the study of the nature, life and actions of micro-organisms.

Diagnostic work in microbiology is carried out in the United Kingdom mainly in the pathology departments of National Health Service hospitals and the regional laboratories of the Public Health Laboratory Service (PHLS). Private hospitals often have a small laboratory for carrying out routine diagnostic tests, the more sophisticated tests are contracted out. All laboratory provision for microbiological purposes is controlled by the Department of Health and Social Security (DHSS).

In addition to the regional laboratories, PHLS is also responsible for the Centre for Applied Microbiological Research (CAMR) which is the designated centre for handling dangerous pathogens. The work of CAMR embraces research into diagnostic techniques, development of vaccines and therapeutic products, and the manufacture of these items under pharmaceutical conditions.

The need to handle dangerous pathogenic organisms in the course of diagnostic research procedures or in the processes of manufacture has concentrated attention on the environment in which such procedures take place. Containment of infection is essential in order to protect the population, both human and animal, at large and the workers within the laboratory. Failure to achieve this would place laboratory workers at risk and could cause serious epidemic disease.

To achieve a balance between user requirements, the advancement of science, and the protection of the human and animal population, three aspects of laboratory safety must be considered:

(a) the layout, construction, equipping and environmental requirements of the laboratory and its associated spaces;
(b) the application of recognised safety measures in the handling of dangerous pathogens; and
(c) the training and supervision of laboratory staff.

THEORY OF DEFENCE AND DISCIPLINE

Containment of infection involves the provision of a series of barriers around the infected material. The number and type of barriers required depends on the level of hazard presented by the organism in question. That is, the more dangerous the organism, the more stringent are the defence regulations.

Figure 31.1 demonstrates the barriers which can be erected around the infected material. Starting at the centre and working outwards there are four: a safety cabinet within which the material is handled; the laboratory and its

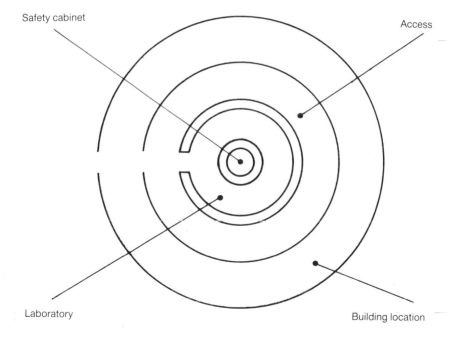

Fig. 31.1 — Laboratory defence barriers for the containment of infection

ventilation and drainage; the access and approach to the laboratory, perhaps via changing and gowning rooms, and its location within a building; the location of the building which houses the laboratory. For example, a building containing laboratories handling dangerous pathogens should not be located near densely populated areas.

The defence provided by these physical barriers is reinforced by disciplinary barriers including staff training and supervision, health supervision and the use of protective clothing. Additionally there should be strictly defined procedures for the transport, reception and handling of specimens and for the disposal or reprocessing of waste material and used equipment, and strict control of access to laboratories and to the building itself.

CLASSIFICATION OF ORGANISMS

Pathogenic organisms are classified according to the level of hazard which they present to people in general and to laboratory workers in particular. There are several systems of classification. However, none of these systems is quite straightforward. For instance, hazard to the individual workers and to the community is not always the same, and the level of hazard may depend on different factors. Furthermore, the various classifications are not exactly comparable one to another. It is necessary therefore in any specific situation to be quite clear about the organisms involved, the quantity of material to be used, the procedures to be carried out, and the nationally accepted classification system.

In the United Kingdom there are four categories [1,2]. *Category A* contains organisms that are extremely hazardous to laboratory workers, some of which may cause serious epidemic disease if they escape from the laboratory. These require very stringent conditions for their containment, and fortunately they are very rarely encountered in hospital clinical laboratories. *Category B1* organisms are a special hazard to laboratory workers but not to other people outside the laboratory; that is, epidemics are unlikely to occur if a laboratory worker becomes infected. Special accommodation and conditions for containment must be provided for handling these organisms. *Category B2* was created especially for hepatitis, which requires special conditions for containment but not special accommodation. *Category C* includes any organisms that are not listed in Categories A, B1 and B2, and that offer no special potential hazard to laboratory workers provided that a high standard of microbiological technique and safety is observed.

In the United States micro-organisms are classified according to hazard by a numerical system[3]. *Class 1* contains 'harmless' microbes; *Class 2* corresponds roughly to the United Kingdom's Category C; *Class 3* to Category B1; and *Class 4* to Category A. A similar system[4], proposed recently by a World Health Organisation (WHO) Working Group, defines Risk Groups 1 – IV in terms of hazard to the individual worker and to the community.

The Health and Safety Commission (HSC) in the United Kingdom has proposed yet another classification[5] in which organisms are listed in Schedules 1 and 2 roughly comparable but by no means equivalent to Categories A and B respectively. The various systems of classification are shown in Table 31.1.

Table 31.1

The classification of organisms

RISK	GODBER[1]	HOWIE[2]	HSC[5]	USA[3]	WHO[4]
High	A	A	Schedule 1	4	Risk Group IV
		B1			
	B		Schedule 2	3	Risk Group III
		B2			
	C			2	Risk Group II
Low				1	Risk Group I

MICROBIOLOGICAL SAFETY CABINETS

The first line of defence against pathogens or dangerous organisms is the safety cabinet. In the United Kingdom it is now the practice to provide a microbiological safety cabinet for the express purpose of providing protection to the user and the environment from the hazards associated with the handling of infected and other biological material. The safety cabinet as designed is not for the use of radioactive, toxic or corrosive substances. The exhaust from cabinets in all cases must be filtered on leaving the cabinet before discharging into the atmosphere. It must be stressed that a fume cupboard is not a safety cabinet in this context and must never be installed in a Category A laboratory. The definitions of safety cabinets are contained in a British Standard Specification[6] and guidance on their choice, installation, maintenance and use in a booklet issued by DHSS[7].

In the United Kingdom only two of the safety cabinets described in BS 5726[6] are accepted for use in microbiological research. These are Class I and Class III. Class I cabinets draw air through an opening from the room, for work on materials suspected of contamination with, at worst, B1 pathogens. If there is thought to be any risk of infectious particles being generated, it is not accepted safety procedure to carry out these manipulations within a safety cabinet. The use of Class II cabinets for manipulation involving pathogens has not become an accepted practice as recirculation of air into the room from the infected material is a risk. The use of Class II cabinets has been restricted to such operations as medicine preparation to ensure sterile conditions of the medicine.

The design characteristics of three types of safety cabinet are illustrated in Fig. 31.2.

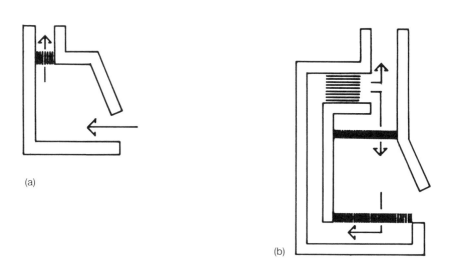

(a)

(b)

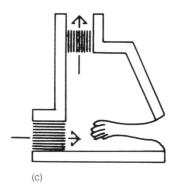

(c)

Fig. 31.2 – Design characteristics of microbiological safety cabinets: (a) Class 1, (b) Class II, (c) Class III

MICROBIOLOGICAL LABORATORIES

The second line of defence is the design of the laboratory itself, involving planning aspects, construction and finishes, fittings, and the control of the environment with its mechanical services.

Function of a laboratory

A laboratory is a place where scientific experiments and investigations are carried out. Microbiology laboratories may involve the functions of diagnosis, teaching, research or production. The various forms of diagnostic work will normally deal with organisms up to and including Category B1 and such work is usually carried out in pathology laboratories. For Category B2 organisms, special provision has to be made within the laboratory, while Category A organisms require a wholly dedicated environment. Teaching will normally be concerned with B1 organisms and rarely with B2. The purpose of these laboratories is instruction in manipulation, various techniques and the study of organisms themselves.

An environment may also be devoted to production, for example of vaccines, at various scales. At the smallest scale, production for clinical trials may take place in a laboratory. There is also need for production on a limited scale for specific requirements. This falls short of the production processes for quantities which are commercially viable.

Planning principles

There are two basic conditions to be taken into consideration when planning the laboratory. The 'clean' condition where the sample is protected from contamination by the laboratory workers, and the 'toxic' condition where the laboratory worker is protected from infection by the sample.

In the clean condition the room relationships are such that the individual enters from the outside, showers to remove external contamination, dresses in sterile clothing, then enters the clean area. On leaving, no particular precautions are necessary because the laboratory worker is simply returning to the relatively less clean environment of the outside world. The relationship of the rooms is shown in Fig. 31.3.

In the toxic condition the planning relationships are reversed. First, the individual enters a changing area to undress and put on protective clothing before entering the toxic area. On leaving the worker has to undress, then shower to remove any possible contamination before dressing to return to the relatively cleaner environment outside. In both cases, either the clean room or the toxic room is the area of containment. The relationship of rooms is shown in Fig. 31.4.

The most critical factor in the provision of a suitable laboratory environment is the design of the mechanical ventilation. The required containment is achieved firstly by the siting of the safety cabinet in relation to the means of ventilation in the case of Class 1 cabinets, and secondly by a cascade system of positive pressures which follow the planning relationships of the laboratory suite.

In the clean condition, the pressures increase room by room from the ambient level outside to the most positive pressure of the area of containment so as to hold back any possible air-borne contamination. In the toxic condition, the

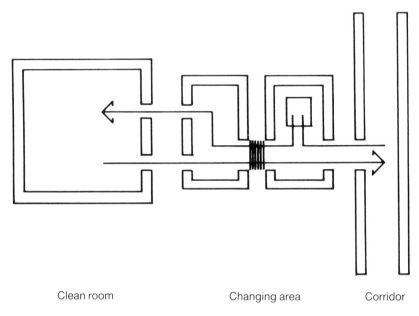

Clean room   Changing area   Corridor

Fig. 31.3 – Basic relationships to be considered in the planning of 'clean' rooms

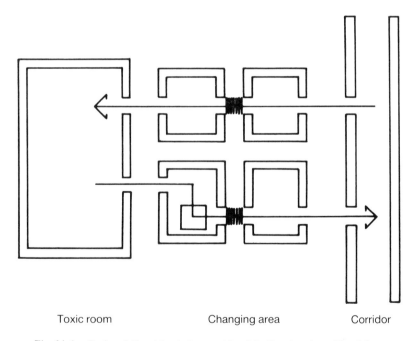

Toxic room   Changing area   Corridor

Fig. 31.4 – Basic relationships to be considered in the planning of 'toxic' rooms

reverse is true, with the pressures being decreased step by step towards the area of toxic containment.

The air cleanliness level within the area of containment is achieved by mechanical filtration using high efficiency particulate air filters. In addition to the air filtration, the air change rate and degree of recirculation is important. In the working area a minimum of 20 air changes per hour is required as defined in the relevant British Standard Specification[8].

Construction and finishes

The British Standard[8] requires internal surfaces to be smooth, impervious and free from cracks and cavities. They should not shed or accumulate particulate matter, and should permit repeated applications of cleaning and disinfecting agents. To facilitate cleaning there should be no recesses or projecting ledges, and coving should be provided between walls, floor and ceiling and be continuous. Where sheet materials are used, joints should be welded and flush. Windows should be non-opening, sealed and flush. All permanent pipes and cables should be boxed in and ducting manufactured from cohesive, resistant materials.

A number of methods may be used to meet these requirements. The traditional walling techniques use plaster on brickwork covered with vinyl sheeting or proprietary paint finishes, or alternatively, plasterboard on studding with similar surface finishes. There are also a number of proprietary systems offering metal skin partitioning with stove enamelled paint finish or plastics coating. Other specialist systems use plastics laminate on studding, polyvinyl chloride on stud partitioning, and moulded and modular glass-reinforced plastics panels.

Choice of the system will have to take into account a number of factors such as inherent stability to avoid cracking, the accommodation of ducting to floor level, and any requirement for flexibility and demountability.

SUMMARY

The first issue in the design of microbiology laboratories is to determine the precise function, and the category of organism to be dealt with. From this will follow the system of defence beginning with the type of safety cabinet, the containment room itself and the relationship of associated facilities. The design of the mechanical services and the building finishes are all aimed at maintaining an appropriate level of uncontaminated environment to protect the work and the occupant.

REFERENCES

[1] *Report of Working Party on the Laboratory Use of Dangerous Pathogens (Godber Report),* Command Paper 6054, HMSO, London, 1975.

[2] *Code of Practice for the Prevention of Infection in Post Mortem Rooms,* Department of Health and Social Security, HMSO, London, 1978.

[3] *Classification of Etiological Agents on the Basis of Hazards,* Atlanta Centre for Disease Control, US Department of Health, Education and Welfare, 1974.

[4] Safety Measures in Microbiology: Minimum Standards of Laboratory Safety, World Health Organisation, *Weekly Epidemiol. Rec.,* 1979, **54**, 337.

[5] *Guide to the Health and Safety (Dangerous Pathogens) Regulations 1981,* (HS) (R) (12), Health and Safety Executive, HMSO, London, 1981.

[6] BS 5726: 1979, *Specification for Microbiological Safety Cabinets,* British Standards Institution, London, 1982.

[7] Health Equipment Information No. 86: *69/80 Microbiological Safety Cabinets; Recommendations concerning their choice, installation, routine maintenance and use,* DHSS, Health Service Supply Branch, June 1980.

[8] BS 5295: 1976, Parts 1 to 3, *Environmental Cleanliness in Confined Spaces,* British Standards Institution, London, 1976.

32

Microbiology laboratories – relating design to risk in the light of legal requirements

M. S. Griffiths
Health and Safety Executive, Edinburgh[†]

INTRODUCTION

This chapter discusses the legal requirements dealing with the design of micro-biology laboratories. Although it relates particularly to microbiology laboratories, the way in which this problem is approached has a general application to other laboratories. The legal requirements discussed are not those associated with, for example, fire precautions, nor to any great extent with those dealing with discharge of liquid waste which could be dealt with under Pollution Acts. They deal with the Health and Safety at Work, etc. Act 1974 (HSW Act), covering the United Kingdom, which is concerned with the health and safety, not only of people working within the laboratory, but also of people outside who may be affected by the activities that go on within those laboratories.

The HSW Act converted the common law duty of 'reasonable care' into a statutory duty under the Act. It applies to employers and other people engaged in work activities. The central theme of the Act is that one should do all that is 'reasonably practicable' to ensure the safety and health of people who may be affected by those work activities. There is the clear implication that the risk perceived or expected should be balanced against the cost of remedying that risk. If a risk is low then the cost of remedy should be concomitantly low, while if high the cost of the remedy may be very expensive indeed. The cost need not simply be financial, it could be effort, inconvenience and so on. The balancing act produced by this general statement is not defined in the Act but is derived from decisions taken in the courts.

---

† Now with Scottish and Newcastle Breweries plc, Edinburgh.

THE DESIGN AND OPERATION OF LABORATORIES

The Act states that in conducting an undertaking, the employer must maintain his place of work in a safe condition. He must provide and maintain a safe working environment and must not expose others to the risks deriving from his work activities. These three statements clearly have implications for the design, operation and use of laboratories. It is very difficult to maintain a place of work in a safe condition if the design of the laboratory is inherently unsafe. It is impossible to provide and maintain a safe working environment if the ventilation and air handling arrangements of a laboratory are faulty. Where the risks are high, unless proper provision is made for the filtration of air and treatment of liquid waste as they leave the building, people outside the building may be exposed to risks from the work activity. Implicit in the Act is a requirement to anticipate the risks and to design the laboratories accordingly.

The HSW Act also carries with it a novel experience for most laboratories, that of external inspection by the Health and Safety Executive (HSE). One purpose of this external inspection is to ensure that the occupiers of buildings assess the risks that might be anticipated and design the building accordingly. It is not uncommon for plans and ideas to be submitted to HSE for comment at a fairly early stage in design; this is welcomed by specialist inspectors. This can be very helpful because HSE can often make comments which are practical and save money. There are of course occasions on which costs rise in order to comply with safety requirements. A main advantage of this system is that it leaves scope for innovation and experiment and does not generate a set of immutable regulations from which there is no retreat or advance. Such regulations often have an inherent disadvantage of specifying a single solution to a particular problem. The 'reasonably practicable' approach allows for experiment and change but always, quite rightly, places the responsiblity for achieving at least equivalent standards of safety to those existing already upon the designers and users of the building. This advantage of innovation is useful to designers wishing to make changes; but, from the user's point of view, what is required really needs to be known more clearly.

PHYSICAL CONTAINMENT

Central to the theme of the HSW Act is that the physical containment provided should be sufficient to contain the perceived risk. It is therefore essential to have a good idea of what the level of risk is.

Figure 32.1 portrays diagrammatically a guide to the levels of infectious risk that can be anticipated. School laboratories often handle material of negligible hazard, such as yeast for making bread, yoghurt bacteria and perhaps organisms isolated from the environment which are not pathogens. Occasionally they will encounter low grade pathogens. Similarly, Colleges of Further Education tend to

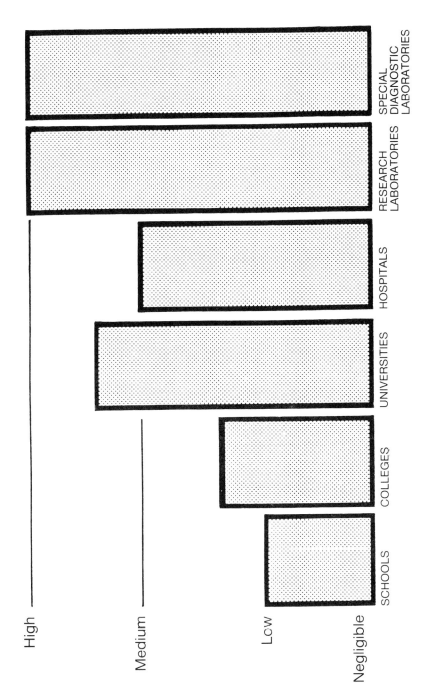

Fig. 32.1 — Levels of risk involved in the activities of various establishments

stay in the low-to-negligible risk category but some look, for example, for food-poisoning micro-organisms. Universities, on the other hand, quite commonly wish to handle much more dangerous micro-organisms, for example the causative agents of tuberculosis, typhoid or brucellosis. Also universities frequently wish to handle them in conditions where the quantities or volumes involved are greater than those normally found in an equivalent hospital laboratory dealing with the same sort of organisms. It is for this reason that the risk rating has been pitched at a slightly higher level than that anticipated in hospitals. Some evidence exists which tends to suggest that the possibility of infection is higher in university and research laboratories than in hospital laboratories[1].

Research laboratories handle anything from totally innocuous organisms to the most dangerous group of organisms, for example, Lassa Fever virus, Ebola virus, etc. Finally, there is a small group of very specialised diagnostic laboratories which handle specimens from patients who might be infected with dangerous organisms. These high risk organisms, and there are very few of them, are designated Category A, the medium risk organisms Category B, and the remainder, common pathogens responsible for the majority of infections, Category C. This categorisation allows the designer and user to pitch a requirement for containment against each of these categories. These requirements, for a variety of reasons, have developed from guidance given by various official government bodies. It must be emphasised that this guidance is not a statutory requirement under the HSW Act but, in practice, is often regarded as 'reasonably practicable'. Since laboratories may be built for one government department or another, the requirements have, in effect, a very persuasive influence on the final design, but again it should be emphasised that they are not legal requirements in the sense of Regulations. There is some element of discretion allowed within the guidelines.

PUBLISHED STANDARDS

A Working Party, chaired by Sir James Howie, produced a report which discussed not only the working practices but to some extent the design features of hospital laboratories dealing with organisms up to Category B. This Report and the Code of Practice which arose from it[2] resulted in an amendment by the Department of Health and Social Security to their *Health Building Note, No. 15*[3] which applies to the construction of hospital laboratories.

A United Kingdom expert committee, the Advisory Committee on Dangerous Pathogens (ACDP), has been formed recently with the general overall task of categorising the hazards and risks associated with infectious micro-organisms. The Committee's classification scheme is likely to have four categories of risk, Category 4 being the highest (equivalent to Category A) and Category 1 being organisms not liable to cause infections. It is currently revising the containment standards for laboratories and is also considering what levels of containment are required for work with infected animals. In due course it will attempt to rationalise any anomalies in infectious hazards and bring together risk assessment to produce a coherent single categorisation scheme.

A second expert group formed several years ago is the Genetic Manipulation Advisory Group (GMAG). This was to deal with the conjectural hazards that various scientists thought might be associated with the handling of genetically engineered organisms. These are organisms in which a foreign DNA, for example human DNA, is inserted into a bacterium and the bacterium is grown, producing the foreign protein that that DNA specified. Clearly if the DNA specifies something rather toxic then a large amount of toxic material could be produced. In fact, the actual risks seem to be substantially lower than the anticipated risks. There have been, for example, no recorded cases of infection from a genetically engineered organism. This has led to a general lowering of containment requirements for experiments as the perception of hazard has fallen and more precise estimates of risk have been made. On the other hand, some risks are foreseeable and a Code of Practice on containment levels[4] has been drawn up recently to cope with these possible risks. Four levels of containment have been specified, Level IV being the highest and Level I the lowest. Level IV is equivalent to Category A dealing with manipulated organisms containing nucleic acids from these exotic viruses. Level I is roughly equivalent to Category C of the Howie and ACDP systems. A notional Level 0, or good microbiological practice, has also been introduced.

These groups and committees produce guidance which, at the moment, is still in a state of flux. There have been substantial changes in microbiological laboratories in the last two or three years and will continue perhaps for another year or two but there are signs that there will be a coalescence of standards, with perhaps eventually a unified graded containment requirement matching precautions to risk. Fig. 32.2 summarises the current position and tries to show how there is a basis for a coherent and rational approach which will relate the perceived or actual risks to the required level of containment. For clarity, detailed differences have been deliberately left out and only the basic principles where there is common agreement have been shown.

GENERAL DESIGN FEATURES

Figure 32.2 shows that there are certain elements of laboratory design which are held in common by all laboratories. The floor should be impermeable, for example of seam welded polyvinylchloride which, if possible, is coved at the edges. Walls should be washable and, in the higher-grade laboratories, plastics finishes are often the norm, since these allow for shrinkage and other movement in the building. This requirement is not so essential at lower levels. It is essential that bench surfaces should be smooth and impermeable and they can be of wood, laminate or stainless steel. There is a tendency to provide more expensive finishes as the containment level increases. Separate hand washing facilities while

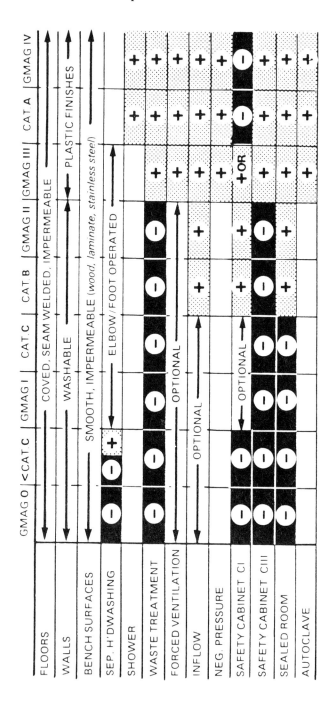

Fig. 32.2 – Relationship of risk to design

not essential at the very low levels of risk are desirable, but once there is work associated with any reasonable risk, there is a requirement to provide separate hand washing sinks which have non-touch type controls. At higher levels there is a need to provide a shower. Waste treatment, either by chemically inactivating the effluent with disinfectant or by heat treatment, becomes an essential at Category 3 or 4 and Category A.

VENTILATION

Ventilation is a vital element in the laboratory design since many laboratory acquired infections occur in the absence of an identifiable accident. It is commonly assumed that aerosols are responsible for many of these infections[1]. Aerosols may remain suspended in the air for long periods of time and could be distributed by badly designed ventilation systems to other parts of the building. There is no requirement to have a forced ventilation until Level III at genetic manipulation or Category A is reached. However, since many hospitals or universities have an air conditioning system for the whole building, it is important that the ventilation requirements of the Category B and GMAG II[4] laboratories are compatible with the ventilation applied to the rest of the building. It should be emphasised that it is at this point that many design faults occur and problems most frequently arise. As previously stated, it is critical that in the event of a hazard being generated within a room, the design of the building is such that dissemination of that hazard throughout the rest of the building does not occur. There is, therefore, a great advantage in having a net directional inflow of air into all rooms handling micro-organisms, but this advantage becomes an obligation at Category B level, and even more sophisticated requirements apply at Category A level. For Category A and Levels III and IV, there is a requirement for a minimum of 7 mm water gauge negative pressure to be maintained within the room.

SAFETY CABINETS

Primary containment in laboratories is provided by a safety cabinet[5]. Typically at Category B level this is a Class 1 open-fronted cabinet which has a limited capacity for containment, but is considered sufficient for this level of risk. At GMAG Level III it is optional whether this or a totally enclosed glove box, known as a Class III Cabinet, is provided. Both types of cabinet, should be ducted into the room extract system or to atmosphere via the integral *HEPA* filters. This minimises the risk in the event of filter failure and allows the safe discharge of the irritant and toxic formaldehyde vapour used for regular disinfection. At

Category A level and Level 4, totally enclosed glove boxes become essential. For Category B and above, the room should be sealable to allow fumigation in the event of an accident or catastrophe, and for routine maintenance. Finally, an autoclave should be provided at high levels of containment, for Category A this is a double-ended autoclave.

The effect, in practice, of these increasing requirements with regard to levels of risk is to create two levels of containment – up to Category B or Level 2 there is only primary containment. Hazardous activities are conducted within the safety cabinet which provides a single barrier, but because this is inside a room which itself will be under a net directional inflow of air, some degree of secondary containment may exist. The simplest arrangment, which may not always be adequate for comfort conditions, is a room with no air conditioning in which a cabinet is installed. Make-up air for the cabinet comes from surrounding rooms and corridors. Air passes into the room through transfer grilles, into the cabinet and is filtered before discharge to atmosphere. For air conditioned rooms more complicated arrangements have to be introduced to ensure that in the event of failure of the extract fan the room does not become pressurised.

At Category A or Level 4 there are two barriers, an air extract system for safety cabinets with filtration for cabinets at each point, and a separate extract system for room air through *HEPA* filters, there is also an input of filtered air. Input air must be filtered to allow the room to be sealed in the event of a power failure. At this level of containment standby generators are also required. In addition effluent waste must be disinfected either chemically or by heating. Waste from the autoclave, shower and sink is diverted to treatment tanks. There is a primary barrier of the cabinet and a secondary barrier of the room. In the event of a failure of the primary barrier the secondary barrier should be designed to contain any release and prevent it escaping to the environment. This limits exposure to those working in the room who can then be placed under strict medical surveillance.

LABORATORY LAYOUT

A plan of a typical Category A laboratory is shown in Fig. 32.3. The Class III cabinets (1–5) are interconnected with each other and with the autoclave (6). These together with the *HEPA* filtered air extract system fitted to them present the primary barrier in which all 'open' work with dangerous pathogens takes place. The secondary barrier consists of the laboratory and ante-rooms. A progressive increase in negative pressure exists, the clean change room being only slightly negative and the laboratory higher (minus 7 mm water gauge with respect to the lobby). All *HEPA* filters are ceiling or wall mounted which ensures 'clean' ductwork. All major services are accessible in a roof void above' the laboratory. Effluent from sinks, shower and autoclave is collected to a heat treatment tank (not shown in Fig. 32.3).

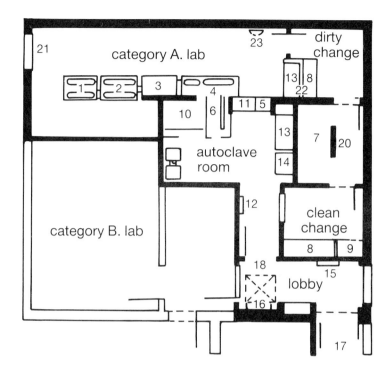

Key: 1 & 3 work cabinets; 2 centrifuge and microscope cabinet; 4 autoclave cabinet; 5 dunk tank; 6 double-ended, high security autoclave; 7 shower (with magic eye); 8 seats; 9 lockers; 10 autoclave compartment; access to controls; 11 window; 12 switches (air handling unit); 13 sink; 14 washing machine/tumbler drier; 15 recorders, manometer, indicators, alarms, etc.; 16 ladder to plant room (above); 17 hazard warning sign; 18 'Work in Progress' sign; 19 refrigerator and deep freeze; 20 shower controls; 21 emergency exit; 22 partition (hall height); 23 convex mirror.

Fig. 32.3 – Layout of a typical Category A microbiology laboratory.

SUMMARY

High containment microbiology laboratories are highly specialised and require a high level of maintenance both from engineers and laboratory staff. They are neither flexible nor adaptable and run counter to the general approach adopted in some other chapters. Flexibility has to suffer where containment requirements become paramount. Where there is flexibility it is in the application of present standards, which provide scope for innovation and development.

ACKNOWLEDGEMENTS

I am grateful to Kelly/Williams Partnership for supplying the plan of the Category A laboratory (Fig. 32.3) and to Dr Gibson, of the Public Health Laboratory Service, for permission to use it.

REFERENCES

[1] R. M. Pike, *Health Lab. Sci.,* 1976, **15**, 105.
[2] *Code of Practice for the Prevention of Infection in Clinical Laboratories and Post Mortem Rooms,* Department of Health and Social Security, HMSO, London, 1978.
[3] *Health Building Note No. 15: Pathology Departments,* Department of Health and Social Security, HMSO, London, 1981.
[4] *Code of Practice for Genetic Manipulation Containment Facilities,* Genetic Manipulation Advisory Group Note 15, Medical Research Council, London, 1981.
[5] BS 5726: 1979, *Specification for Microbiological Safety Cabinets,* British Standards Institution, London, 1982.

33

The design of a new radiochemical laboratory complex

A. G. Lewis

Amersham International plc, Cardiff

INTRODUCTION

In 1940, the founder of Amersham International (formerly The Radiochemical Centre), Dr Patrick Grove, undertook some work on the refining of radium to prepare luminous paint for use in instruments in ships and aircraft. This work was later extended to the encapsulation of radium into needles for use in the treatment of tumours, and eventually a centre was set up to supply a service for radioactive materials, and to exploit the commercial possibilities throughout the world, where possible. This was the beginning of a medical isotope service in the United Kingdom.

Towards the end of the Second World War, the world's first nuclear pile was built in America, and the first controlled nuclear reaction initiated. The possibility of introducing stable chemicals into an atomic reactor and thus creating a range of artificial radioisotopes opened up new horizons. Eventually, after much debate as to who should be responsible for the developments of radioactive materials in Britain, it was decided that all the preparation and commercial exploitation of this work would be carried out by the Radiochemical Centre at Amersham.

In the post-war years, the Company gradually developed its business into four distinct areas, each one associated with a different aspect of science and hence a different market. The aim of the company at that time was to

provide a dependable service of supply of these important substances for use in medicine, research and industry, wherever they may be needed at home or abroad, as a commercial enterprise.

The four areas of business were:

(1) *Radio-pharmaceuticals,* which are directly administered to human beings by oral or intravenous methods, and are used in the diagnosis of organ malfunctions within the body.

(2) *Clinical reagents,* which are used in kit form to examine the levels of minute amounts of substances such as hormones and antibodies in body fluid samples.

(3) *Labelled organic chemicals,* which are used as analytical tools in most chemical and biochemical establishments. The radioactive label enables the organic compound to be traced in a research environment or in living systems.

(4) *Radioactive sources,* which are used medically for radiotherapy and in industry for measuring the thickness of materials such as paper and steel, industrial radiography, sterilisation, smoke detection and a multitude of other purposes.

THE NEED FOR EXTRA FACILITIES

The Company's activities expanded steadily until in 1973 it became apparent that the existing site of approximately 7 hectares at Amersham would not contain the planned future expansion of the Company. There were very clear ideas within the Company about the new development; a number of the desirable features are listed below:

(a) A quality site, which would provide a pleasant working environment and accommodate site requirements such as car-parking and recreational facilities.

(b) Good motorway communications with the parent site and with the major London Airport (Heathrow).

(c) Access to a university and to colleges with good training facilities.

(d) Access to an area of high population offering good opportunities for recruitment at all levels.

(e) A site which would permit expansion over the next ten years to a size of approximately 1000 staff — a level perceived by the company to be an optimum on one site.

(f) A site which would be sufficiently remote from Amersham for it to be effectively independent, with its own supporting technical and administrative service departments.

In 1974 the Company decided to purchase 12 hectares of land on the northern outskirts of Cardiff to develop a second major site to accommodate 1000 staff. The plan was to establish roughly half the Company's technical work at Cardiff, i.e. two of the four business areas described earlier. These were the departments producing clinical diagnostic kits and labelled organic chemicals.

It was decided to complete the work in two phases which are referred to as Contract 'A' and Contract 'B'. The objective of Contract 'A' was to provide an engineering building which would also house stores, despatch operations and other support services, a power house to supply some of the centralised services such as steam and compressed air, and an effluent disposal area where liquid and solid waste could be accumulated before proper disposal. This work began in 1976 and the Contract 'A' buildings were occupied in November 1977. The completion of these facilities was important, as it provided an on-site base for the project team during the later stages of planning, and the construction of Contract 'B'. It enabled small numbers of staff to be recruited in all functions, though most of the technical laboratory staff were sent to Amersham for extended training periods. The building was also used to house approximately £2 million worth of plant and equipment purchased by the Company for installation in the laboratory buildings on the completion of the whole contract. All of this equipment, available well in advance of its required date, was rapidly installed in the laboratories as soon as access was allowed by the main contractor. This was a key part of the plan to enable quick movement past a commissioning phase and into normal production operations.

Construction work on the Contract 'B' buildings commenced in August 1977, and was intended to be a two-year contract. One of the laboratory buildings, Medical Products, was to be finished in May 1979, with the whole of the rest of the site being completed by August 1979. In the event this plan was extremely optimistic and final handover of the whole site was approximately 15 months late. However, during this period of delay, it was possible to take over various parts of the complex and use them, even though the Company was not in full possession of the whole site.

Perhaps at this stage, it is appropriate to mention briefly some of the problems of relocation to a completely green-field site, and also some of the technical problems faced by the design team. The first major problem was the withdrawal of planning permission by the Secretary of State in 1974 following local opposition to the development of the Forest Farm site. This resulted in a public inquiry, and effectively delayed progress on the project by approximately 14 months.

Although the technical specification for the laboratories was precise, the architect was given considerable discretion in the layout of the site and particularly in the use of building materials. Two of the chosen materials, the glass-fibre reinforced polymer panels for the external cladding and the reflective glass for windows and curtain-walling, both created some technical and co-ordination problems. The sheer complexity of the services in the laboratory areas, particularly the ventilation systems, caused further enormous coordination difficulties. Resolution of these problems required very intensive discussions between the architect, the consultants and the main contractor throughout the contract.

Of course, although not directly relevant to this chapter, one of the major problems of relocation involved personnel — both the transfer of about 100 existing staff and the recruitment locally of approximately 350 staff. Even considering the delays in occupying the building, the transfer and recruitment of staff was accomplished remarkably smoothly. The training schedule, particularly for technical laboratory staff, was intensive and required a major effort by existing staff at Amersham.

THE LABORATORIES

The two laboratories constructed at Cardiff are intended for two completely separate and different production processes. The Medical Products building is used for the production of a range of approximately 40 diagnostic kits for use in the treatment of thyroid, obstetrics/gynaecology and haematology/drugs disorders. The main isotope used is iodine-125.

The Chemical Products building houses two separate departments responsible for the development and manufacture of a wide range of approximately 2000 organic compounds, which are labelled with either tritium or carbon-14.

It was decided to build two laboratories of the same general external shape and to separate them by a building housing the major laboratory services, particularly the ventilation systems. The technical buildings were built near the centre of the site, with the engineering complex at one end and the offices and amenities at the other. All buildings are linked by a spine corridor, which at ground level permits the movement of people and paper between buildings; at basement level, the spine duct carries all electrical and mechanical services, as well as telecommunications and computer links.

Each laboratory is approximately 50 m square, giving a total floor area of rather more than 5000 m². Each floor of the laboratory is divided by a central corridor; emergency exits are mainly external to the building envelope in order to maximise the use of laboratory space. One of the corridor walls acts as a one-hour fire partition, and so each laboratory is effectively split into four separate fire-distinct areas. Fire compartmentation is an important part of the internal control and planning procedures. Ancillary laboratory space, cold rooms, and offices are generally located on either side of the central corridor, but are also used to sub-divide the large open laboratory spaces into smaller areas where appropriate. This arrangement creates a very deep laboratory working space, which allows most of the perimeter to be kept free of plant and equipment. Work stations, which include a range of ventilated enclosures, open work benches, writing areas, and zones for free-standing plant and equipment, are all located in the deep laboratory space, leaving the perimeter for communication purposes. With the exception of drainage, all mechanical and electrical services are installed in the ceiling void above the working area. Access to these services is through a fully demountable ceiling, into which have been integrated the plenum and extract ventilation grilles, the fluorescent lights and the various work stations which reach ceiling level.

The most fundamental service in the laboratories is the ventilation system. Because of the nature of the work, the laboratories themselves are held at a slight negative pressure, which is achieved by accurate balancing of the plenum and extract systems, to create a differential of approximately 5 per cent. None of the windows in the laboratory areas can be opened, although they are tinted and double-glazed to improve environmental conditions. The plenum system brings filtered, fresh air into the working areas at a controlled temperature through a series of centrifugal fans located in the central Air Handling building. Air is ducted into the laboratories in galvanised mild steel ducts, which divide to supply individual quadrants. The extract ductwork is much more extensive and complex, particularly in the Chemical Products building where there are approximately 400 ventilated enclosures, compared with only 40 in the Medical Products building. All extract ductwork is constructed of heavy gauge mild steel, with all-welded seams and flange joints. It is designed to withstand a working pressure of minus 150 mm and a test pressure of minus 200 mm. All surfaces are carefully coated to prevent deterioration due to environmental or chemical attack. Interior surfaces are shot-blasted and then painted with one coat of *Metalcote®*, two coats of *Armourflex®* and a final coat of UPC (Unique Plastic Coating). The exterior surfaces are wire-brushed, and painted with one coat of *Metalcote®* and two coats of *Armourflex®*.

In total, there are eleven extract routes from the Chemical Products building and ten from the Medical Products building. The Chemical Products extract is unfiltered and all air is discharged to atmosphere at high level through two stacks 1.8 m in diameter. The third stack takes extract from the Medical Products laboratory, of which a proportion is routed through filter canisters containing a prefilter/spark arrestor, an absolute filter and a charcoal filter. The total volume of the extracted air is $100 \, \text{m}^3/\text{s}$, and only in the control area of the Medical Products building is any recirculation undertaken; this is done to achieve an accurate constant temperature control of $19 \pm 1°\text{C}$. All extract fans are two-stage axial flow fans, with variable pitch impellors. The extensive runs of ductwork dominate most of the ceiling voids, and all other services have to be integrated with them. Fixing points were cast into the underside of the slab for use as supports for the ductwork, the enclosure head-frames, and all other services: electricity, steam, hot and cold water, chilled water, cooling water, gas, compressed air, nitrogen, argon/methane, oxygen and special gases (argon, helium, hydrogen).

In general, these services are run in ring mains around each floor in the ceiling void. Services are brought to work stations by two different methods:

(a) Ventilated enclosures equipped with fascias containing outlets and valves for the necessary services, which are supplied directly from the ring main through the enclosure head-frame.

(b) Open work-benches supplied through services ducts or 'down-drops'; each service drop is situated at the end of a bench and contains valves behind a maintenance access panel.

The workstations themselves are a dominant feature of the laboratory space. In the Chemical Products building, approximately 400 stainless steel ventilated enclosures are installed. These were designed in-house, to the specific requirements of laboratory staff, but with a major effort to standardise them into a limited number of different types. Apart from very special applications, this was generally achieved. The accuracy achieved by the manufacturer has resulted in a high degree of flexibility and interchangeablity, should this be required at any time. The majority of these enclosures are slit-boxes with Perspex fronts, but there are also glove and tong-boxes in the Medical Products building, together with a range of specially designed enclosures for handling serum, chromatography work and other applications. The design of the work-benches also incorporates a high degree of standardisation. The bench tops are white or grey melamine, supported on prefabricated steel stands of a cantilever design. A standard range of underbench units was designed, incorporating either drawers, storage cupboards, or steel-lined cupboards. The service rail above each bench is in a bright contrasting colour — yellow — which matches the service drops and the cladding of the structural columns. All bench-work finishing materials are tested to Class 0 or Class 1 fire standards. The service rail is also used to support reagent shelves for the storage of chemicals and small pieces of equipment.

The other vital service is drainage. From the beginning, it was intended to make the system as reliable and maintenance-free as possible. The materials selected were polypropylene, using both fusion and mechanical fittings, and borosilicate glass for all floor slab penetrations. Much of the drainage system, with the exception of the complex connections to benches and enclosures, is doubly-contained with the outer containment taken to sumps, which are monitored. Thus any leak in the primary system can be quickly identified. All liquid waste from the laboratories is routed to two 4.55 m^3 collecting tanks in the sub-basement, from where it is pumped through the spine duct in continuous double-contained pipes to the holding tanks in the effluent disposal area.

An extensive programme of equipment inspection and commissioning was undertaken prior to the handover of any area. Considerable efforts were put into the preparation of commissioning schedules, which were rigorously followed by the contractors and the site inspectorate. Detailed specifications were laid down for testing the various services and there is no doubt that this time-consuming and expensive exercise paid handsome dividends in the relatively trouble-free period of post-commissioning.

DESIGN CONCEPTS

Very early in the development of the project, the company appointed a Project Manager, who ultimately became the Site General Manager. The Project Manager was responsible for much of the coordination that took place between the members of the design team — the architect, the consultants and the technical and engineering departments at Amersham. An important factor in the considerations of the design team was that particular products in the Company's range can be superseded quite rapidly, as in some cases the product life cycle may be as short as two to three years. This means that production facilities must be flexible, so that changes can be introduced without undue disruption of existing work.

A typical example of this change in technology is the introduction of the *Amerlex*® range of diagnostic kits, which uses a totally new solid phase separation system for radioimmunoassay, and which was not even at the development stage when the Cardiff site was designed. It is even possible that the Company may enter a new branch of science for a particular application.

During the early stages of project design, serious attempts were made to learn from problems that had existed at the Amersham site. Laboratory staff were invited to specify their requirements and, although some compromise was necessary as the design progressed, many of the individual requirements were provided. The care taken at that stage of the project brought its reward years later at the commissioning stage. Staff were able to move into their working areas, set up their equipment, and start work in a new environment with the absolute minimum of disruption to production. In fact, most customers were never aware of the movement of production facilities from Amersham to Cardiff – deliveries and service were always maintained.

Safety was also a major consideration, both for employees and members of the public. Careful liaison took place between the Company and various official bodies – Nuclear Installations Inspectorate, Ministry of Agriculture, Fisheries and Food, Department of Environment Radiochemical Inspectorate, and Alkali and Clean Air Inspectorate – and the authorising departments – Welsh Office and Welsh Water Authority.

Many pages of safety commentaries were written by Company safety staff, which were discussed and agreed with the various inspectors. Many of these deliberations had a very direct effect on the design of the facilities. Finally, environmental factors were not forgotten. The buildings were designed to be sympathetic with the existing surroundings; the site was originally farmland and is situated in a natural river valley, bounded by a canal and surrounded by trees and open countryside. The overall design, though bold in concept, has blended into the natural surroundings, helped by careful landscaping of the site itself.

CONCLUSION

The last part of the Contract 'B' buildings was taken over from the main contractor towards the end of 1980, approximately fifteen months late. During this delay, it had been possible to negotiate access to parts of the complex so that certain basic operations could take place under Company control. The most significant of these was the transfer of the kit-packing operation from

Amersham at the end of 1979, to its final home on the first floor of the Medical Products building.

In addition, by fitting temporary doors, it was possible to take over individual laboratory areas and fit them out with plant and equipment. As a result, each area was ready to start normal operations in a very short space of time after it was formally handed over by the contractor. Obviously, on such a large contract, there were minor defects to be corrected, and it was necessary to allow tradesmen and inspectors access into working areas long after the handover date. Even so, it was possible to continue normal work whilst these remedial operations were carried out, although in many instances careful supervision was essential.

The site has now been in full operation for some two years, making a major contribution to Company turnover and profitability. Three hundred staff are engaged in laboratory operations, with about another 150 working in technical and administrative support departments. Very few changes have been necessary to the original design during the commissioning and early production phases, but in recent months, several major projects have been initiated, requiring different facilities, plant and equipment. In this technology, this is no more than can be expected.

34

The design of a new glove box facility for handling Plutonium Dioxide powders

R. Taylor
United Kingdom Atomic Energy Authority, Windscale

INTRODUCTION

The Basic Studies Facility for plutonium work at Windscale Nuclear Laboratories is now some 20 years old and requires considerable maintenance effort to ensure the high standards of safety required in such a facility. Recently, a decision was taken to replace the five glove box facility with a modern design (the New Oxide Facility) to ensure effective and safe operation for the next 20 to 25 years.

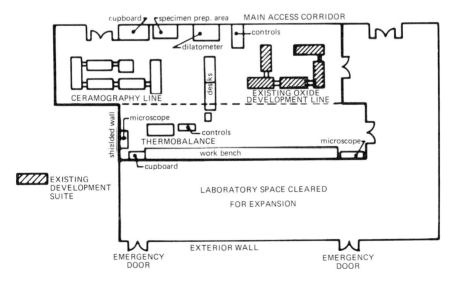

Fig. 34.1 – Existing plutonium oxide line and its surroundings.

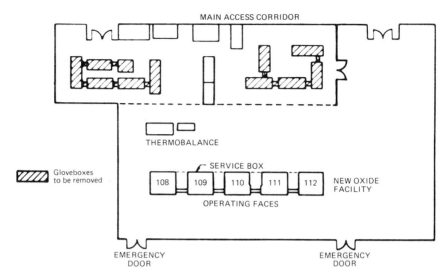

Fig. 34.2 – New plutonium oxide facility and its surroundings.

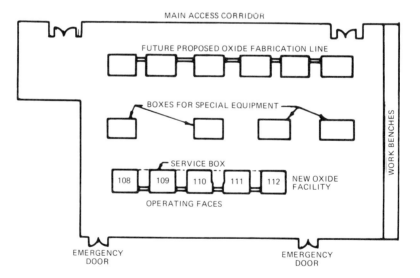

Fig. 34.3 – New plutonium oxide facility and its ultimate surroundings.

The new glove boxes are part of a long-term refurbishment programme for the plutonium laboratories and the sequence of operations is shown in Figs. 34.1–34.3. The existing oxide development line is shown hatched in Fig. 34.1 together with an existing ceramography line which is currently being replaced in another location. The laboratory after installation of the New Oxide Facility is shown in Fig. 34.2 with the old glove boxes for decommissioning and eventual removal shown hatched. The final stage of refurbishment is shown in Fig. 34.3 with the Oxide Facility together with a future oxide fabrication line and individual glove boxes for specialised equipment such as thermobalance, electron microscope and instruments for surface area measurement.

DESIGN PARAMETERS

The main parameters to be considered in the design of a new glove box for plutonium work are the mass of plutonium to be handled, the isotopic composition of the plutonium, the radiological protection problems due to both external (gamma and neutron) and internal radiation (primarily alpha damage to tissue), and the ease of operation and maintenance.

For the New Oxide Facility a series of source descriptions was defined based on past experience of Basic Studies work and the expected work load on this facility over the next few years. Table 34.1 gives details of these source descriptions, expressed in a manner suitable for the subsequent computer calculations of dose rates to be expected in the facility.

Table 34.1

Source descriptions

Case no.	1	2	3
Description	Normal Batch size	Maximum Batch size	PuO_2 Store
Material	$(UPu)O_2$	$(UPu)O_2$	PuO_2
Pu content (%)	35	35	100
Density (g/cm^3)	3.0	3.0	3.0
Mass (g)	100	250	2268
Shape	spherical	spherical	cylindrical
Size	radius = 20 mm	radius = 27 mm	radius = 38 mm length − 171 mm

The external radiation dose to be expected from the plutonium is dependent on the isotopic composition and some indicative data are shown in Table 34.2 where the isotopic compositions have been rounded from those calculated for specific reactor systems, using the FISPIN computer code. The importance of the isotopic composition lies in the differing yields of alpha particles, gamma photons and neutrons from the various isotopes. Plutonium-239, which is the

Table 34.2

Indicative isotope compositions of plutonium

Composition number	1	2	3	4	5
	Atomic (%)				
	3,600 MWd/te	5,000 MWd/te	30,000 MWd/te	Fast	Fast
Type	Magnox	Magnox	CAGR	reactor	reactor
				Inner zone	Outer zone
Isotope					
Pu^{236}	–	–	–	3×10^{-6}	2×10^{-6}
Pu^{238}	0.1	0.2	1	0.3	0.2
Pu^{239}	76	69	40	53	50
Pu^{240}	20	25	41	37	39
Pu^{241}	3	5	10	6	6
Pu^{242}	0.5	1	8	4	5

isotope commonly handled for the last three decades, has only a low gamma yield and a very low spontaneous fission yield. When combined with oxygen to form PuO_2 (or carbon to form PuC) then the production of fast neutrons from the α-n reaction with oxygen (or carbon) has to be taken into account. The energy of the neutron emission is typically close to 2 MeV for spontaneous fission but nearer 5 MeV for those produced by the α-n reaction with oxygen (7 MeV with carbon). If the γ, spontaneous and α-n neutron yields are each taken as unity for plutonium-239 then the yields from other isotopes may be compared.

Plutonium-236 is produced in low yield only in fast reactors via the Np^{237} n−2n Pu^{236} reaction; although its γ yield is some 10^4 times that of Pu^{239} only 10^{-4} per cent has energies greater than 100 keV. The penultimate member of the decay chain is Tl^{209} with decay to Pb^{208} via a 2.6 MeV gamma emission. This may become important in the very long term with repeated recycle round the fast reactor fuel cycle.

Plutonium-238 has a γ output some 10^3 times that of Pu^{239} but only 10^{-4} per cent of these γ photons have energies greater than 100 keV. The fast neutron yields are significantly greater than Pu^{239} by factors of about 10^5 for spontaneous fission and 10^3 for the α-n reaction.

Plutonium-240 has a low γ yield, only about ten times that of Pu^{239}, but the neutron yields are very significant at about 10^5 times that for Pu^{239} for spontaneous fission. This isotope is a major contributor to the fast neutron yield of plutonium produced from high burn-up fuel.

Plutonium-241 is a soft β-emitter with no significant γ emission. The neutron yield is negligible. The prime importance of this isotope is that it is the parent of Am^{241}, and intense γ-emitter. The americium breeds-in after separation of the plutonium in reprocessing plants and its presence is therefore unavoidable. The spectrum from Am^{241} is considerably more energetic than that from the true plutonium isotopes with some 45 per cent of the γ photons between 50 and 100 keV. and 1 per cent greater than 100 keV. Significant yields up to 800 keV are present. The γ dose from Am^{241} dominates that from all other plutonium isotopes within a few months of reprocessing. In addition, because of the relatively short half-life of Am^{241}, the α-n reactions with oxygen or carbon make a major contribution to the fast neutron flux.

Plutonium-242 has a relatively low yield of γ photons and its major importance is associated with the high yield of spontaneous fission neutrons of the order 10^5 times that for Pu^{239}.

From the above it can be seen that, as economic factors drive the thermal reactor fuels to higher burn-up and as the fast reactor system tends towards the equilibrium recycle condition, the isotopic composition of the plutonium tends to become more difficult to handle from the point of view of radiological hazards. In particular, the problem of fast neutron dose rates becomes exceedingly important.

So far discussion has centred on the problems of external dose but it must be recognised that the potential hazard from internal dose still exists. The change in isotopic composition predicted for future high burn-up plutonium has little effect on the allowable Annual Limit of Intake or maximum permissible body burden of plutonium. Therefore, major improvements in the current very high standards of containment are not looked for although, with the 'as low as reasonably achievable' principle now clearly and formally stated, there is still an incentive to achieve the highest standards of containment that are cost effective.

As described above, the external radiation dose from plutonium will increase over the next one to two decades, therefore current and future glove box designs must take account of this slowly changing situation. The design difficulties are compounded by the continuing change in the radiological protection requirements as exemplified by ICRP/26[1], EEC Regulations[2] and the draft of the proposed United Kingdom legislation for ionising radiations[3] which will be introduced in 1985. The last calls for an investigation level of 15 mSv/yr total dose, that is the sum of all external radiation plus all internal radiation should be less than 15 mSv/yr to an occupationally exposed individual. This is a difficult level to achieve with plutonium operations as practised over the last 30 years.

Given the possibility of defining future sources of plutonium, the overall levels of radiological protection required and the source term for the mass and form of plutonium oxide to be handled, then the use of computer codes such as FISPIN and RANKERN allows the calculation of the total shielding requirements to meet specific radiological protection requirements.

THE DESIGN CONCEPT

Since the internal committed dose from plutonium isotopes is not amenable to simple calculation in the way that gamma, or neutron shielding thicknesses may be determined, a basic assumption has been made that an allowance of 5 mSv/yr shall be allocated for the committed internal dose. That is, the 50 yr committed dose, due to one year's intake of plutonium, should be a maximum of 5 mSv. This is equivalent to a dose rate of 0.1 mSv/yr for each of the 50 years following the year of intake. With present standards of glove box design this very low level can be achieved by attention to detail and careful operation. Currently the major risk of incurring internal dose is during maintenance work on extract pipework, filter assemblies and glove box servicing. The new concept of glove box design attempts to eliminate the maintenance risks and therefore aims for a target internal dose significantly less than the 5 mSv assumed. Only by operating experience, over several years, with the new style of glove box combined with full biological monitoring, will it be possible to demonstrate that this allowance is excessive. An increase in external radiation dose above the 10 mSv level may then be considered.

The greater proportion of the γ-radiation from plutonium isotopes is at energies less than 50 keV, therefore thin lead shielding will give a high degree of attenuation. In addition it should be noted that self-shielding, by even small masses of plutonium, is considerable; hence the dose is not mass dependent. In many facilities the greater dose arises from the inevitable very thin layer of dust on the internal surfaces of the glove box.

The neutron shielding problem is more severe since there is no self-shielding and the neutron flux is essentially proportional to the mass of plutonium. With neutron energies in the range 2 to 7 MeV, the neutron flux equivalent to 10 Sv/yr is about 7 $n/cm^2/s$. For thermal neutrons (defined here as less than 10 keV) the equivalent flux is about 200 $n/cm^2/s$. Hence the problem is essentially one of neutron moderation. The most effective common moderators are hydrogen and carbon, usually used in the form of water, paraffin wax or high density polythene. These materials do not differ significantly in the thickness required for a given reduction in neutron flux and in the case of the facility under consideration the shield thickness is in the range 10 to 20 cm, depending on the attenuation required. With a neutron shield of this order, plus the lead gamma shield and the glove box wall thickness, conventional gloved operations become extremely difficult since reach and freedom of movement become very restricted.

A study of the optimum cost of shielding, together with a cost benefit analysis, resulted in a shielding system whereby the gamma dose was reduced to 2 mSv/yr, approximately twice the natural background, by the use of 10 mm of lead shielding. The neutron dose was reduced to 8 mSv/yr using a 110 mm shield thickness of paraffin wax. The total external dose at the outer surface of the glove box was then 10 mSv/yr, the addition of the assumed 5 mSv/yr for internal committed dose then leads to the target of 15 mSv/yr total dose. A set of shielding data for one particular case is shown in Table 34.3 based on the source terms in Tables 34.1 and 34.2 together with the assumption of a 2000 h working year.

Owing to the limitations on reach and mobility imposed by the 130 mm of

Table 34.3

Shielding data for Case 2 (Table 34.1)

Source	Composition number[a]	Shield thickness (mm)		Doses rates outside shield (μSv/yr)		
		Lead	Paraffin wax	Gamma	Neutron	Total
Source 300 mm from inner walls	2	1	0	2.6	2.3	4.9
	3	4	9	0.9	4.0	4.9
	5	3	0	1.4	3.5	4.9
Source close to inner wall	2	10	80	1.0	3.8	4.8
	3	9	110	0.9	4.1	5.0
	5	9	94	0.9	4.0	4.9
Dose rate at surface of source (μSv/yr)	2	–	–	16,000	400	16,400
	3	–	–	38,000	900	38,900
	5	–	–	22,000	600	22 800

[a] See Table 34.2

shielding on the front (operating) face of the glove box a decision was taken not to neutron shield that face but to shield fully the remaining five faces of the box. Then, with the operating faces of the glove boxes turned towards the perimeter walls of the laboratory, the neutron background in the main part of the laboratory would be very low. Further, the front faces of the glove boxes could, easily, be made areas of restricted access and as much instrumentation as possible brought to the rear face of the glove box and outside the neutron shield. An occupancy factor of 50 per cent (1000 h/yr) has been assumed at the working face during the early years of the life of the facility. It is recognised that a thin neutron shield (50 mm) may be required in some five to six years time and provision for this shielding has been made. However, in the longer term it is clear that increased automation and development of artificial arms or mini-robots for use inside the glove boxes will be required, together with closed circuit television, to reduce the occupancy time at the operating face. It is believed that a period of about 10 to 12 years is available before these more sophisticated devices will be required.

The operations glove box will be constructed in stainless steel as a double skin unit. The lead shield and the neutron shield will be totally contained within the double skin structure. The inner skin will be tested to the highest standards of leak tightness so that the leak rate is not greater than 0.05 per cent of the glove box volume per hour at minus 100 mm water gauge. The windows are angled to give an excellent view of the working area and are of minimum size consistent with adequate visibility in the glove box. The primary containment at the window is polycarbonate plastic backed by a 50 mm lead glass gamma shield. All other aspects of the operational box design follow conventional, well proven detail. To improve further the quality of containment and to reduce the risk of potential intake of plutonium by maintenance staff, a service glove box has been designed to be contiguous with the operational glove box. The two glove boxes are placed back-to-back with the neutron and gamma shield wall of the operational glove box between them. The roof of the service box is constructed as a compartmented duct to carry all electrical, instrument and mechanical services to the service box. This system eliminates the maze of overhead pipework and cabling seen in many contemporary plutonium laboratories. Permanently installed pipework is connected between the ends of the service box and the operational glove box and therefore enters the operational glove box in a heavily neutron shielded area. Similarly all electrical and instrument cabling is carried through special multi-pin leak tight plugs and sockets in both service and operational glove box walls. All gases and liquids are filtered at entry and exit from the operational glove box by filters mounted in the service box. Hence, all maintenance work can be carried out under low-activity glove box conditions outside the gamma and neutron shield walls of the operational glove box.

The space between adjacent operational glove boxes is taken up by a 250 mm diameter transfer tunnel linking the boxes and by three heavily shielded, horizontal, storage tubes. Operational controls will be imposed so that any plutonium not in immediate use is always placed in one or other of these shielded stores. The spacing between adjacent operational glove boxes has been increased to 600 mm compared with the conventional 200 to 300 mm. This increase provides additional space in the shielded storage tubes and allows the use of a standard instrument racking system between the glove boxes. The same spacing is also used between adjacent service boxes and in each case the instrument rack faces are continuous with the vertical faces of the operational or service boxes. The whole facility then has a smooth overall envelope which can be easily decontaminated in the unlikely event of a spread of contamination.

Particular attention has been paid to the detail of the mounting of the instrument panels to avoid any crevices or ledges which may entrap radioactive dust likely to become re-suspended at a later date. The instrumentation in the panels between adjacent operational glove boxes is the minimum required by an operator working at the front face of the box. All other instrumentation for operational glove box is provided on the panel between the service boxes. For example, all furnace control equipment and associated electronics is mounted on the service box side of the facility.

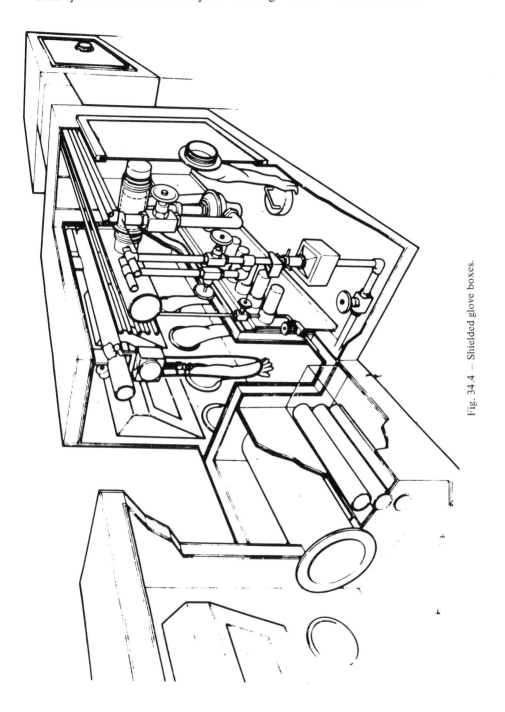

Fig. 34.4 – Shielded glove boxes.

Conventional polyvinylchloride bag posting facilities, including inert gas purged tunnels, are incorporated at each end of the suite of five glove boxes. Provision had been made for conversion to a fully engineered posting system whenever a fully proven and reliable system becomes available. The ends of the facility are fitted with fairings to smooth the outline to the extreme ends of the posting tunnels. These fairings contain the plutonium-in-air sampling systems, the isolation valves and switches for the various services, and additional emergency instrumentation.

A major feature of the design process has been the use of full-scale mock-ups of the glove boxes. Initially these mock-ups were simple *Dexion®* and hardboard structures but eventually they were replaced by full-size wooden structures which have been used to prove the detail of piping runs, valve positions, electrical cabling, etc., and to demonstrate that all equipment can be maintained easily under simulated active conditions. These final mock-up boxes have been invaluable to the design team who have had the benefit of detailed comment from operators, maintenance staff and health physics and safety personnel.

An impression of the new facility is shown in Fig. 34.4 which is a sectioned view from the service box side of the suite. The new facility is under construction and is intended to be operational in 1984.

CONCEPTUAL DESIGN FOR LARGER-SCALE FACILITIES

The concepts described in the earlier part of this chapter will be adequate for the particular scale of operations required in Basic Studies Facilities. However, in the longer-term there will be a requirement to handle significantly larger batches of plutonium dioxide. At these mass levels neutron shielding becomes the dominant design parameter. Neutron shields of some 300 mm will be required and conventional glove box operation becomes totally impracticable even if the extremity (hand) doses could be tolerated. Hence there is a long-term requirement for a new style of plutonium handling facility. One potential concept is briefly described here, and is shown in Fig. 34.5.

The major design consideration is the use of a new type of manipulator suspended from the roof of the neutron shielded cell. The provisional specification for the manipulator is that it shall have a lift of 10 kg at 1 m radius and be capable of reaching every part of the cell. Additionally, the wrist rotation of the manipulator is the prime power source for each item of equipment used in the cell. The manipulator is mounted on an engineered posting system of the double-door type with two other identical posting ports one on each side of the central unit. In the event of breakdown or malfunction, two other manipulators may be plugged in on either side to carry out remedial work, or the central manipulator may be totally withdrawn into its own posting system, disconnected from the cell and moved to a maintenance glove box. This maintenance box would have an identical posting port in the roof and the manipulator could then be extended

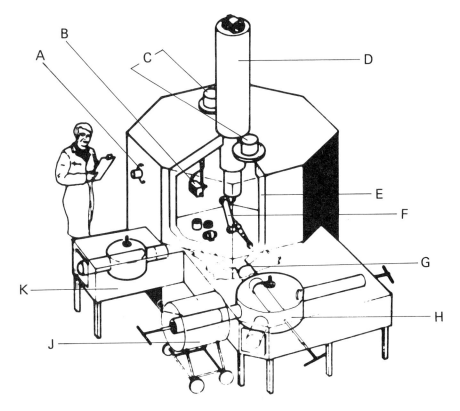

Fig. 34.5 – PORAD facility for plutonium handling. Key: A: Typical lighting position. B: Closed-circuit TV camera. C: Entry ports for extra manipulators. D: Manipulator withdrawal system. E: Neutron/gamma shield wall. F: Manipulator. G: Position of sphincter seals. H: Posting-in system. J: Shielded trolley for incoming materials. K: Active posting system.

into the glove box for hands-on repair work. No massive quantity of plutonium would be associated with the manipulator and therefore only thin lead gamma shielding would be required on the maintenance glove box. A prototype manipulator has been designed and the full-scale wooden mock-up of the operating cell is available for trials. All viewing will be by closed circuit television.

A new system for in-posting using a sphincter seal principle is currently under development for this type of facility. The main design objective here is to ensure easy replacement of the seals as they become worn, with the old seals passing forward into the active volume. *In situ* monitoring, between the sphincter seals, is also being developed.

The sphincter seal system requires container geometry to be well defined in order to obtain the very high standards of leak tightness required in plutonium facilities. In this case a standard cylindrical container will be used. Once inside the cell this container may then be used as an intermediate waste disposal unit through a succession of similar interlinked cells. In the final cell the waste would be transferred into a standard waste disposal drum system and the container removed from the cell via an electro-deplating unit which will remove any contamination. The container would then be re-used for in-posting.

This larger-scale facility is still in the conceptual design stage and full development is not envisaged before 1990.

REFERENCES

[1] *Recommendations of the International Commission on Radiological Protection*, ICRP Publication No. 26, Pergamon, Oxford, 1977.
[2] Council Directive 80/836 (Euratom), *Off. J. Eur. Communities*, **L246**, 15 July 1980, p.1.
[3] *The Ionising Radiation Regulations 198-*, Consultative Document, Health and Safety Commission, HMSO, London, 1982.

35

The design, construction and maintenance of a major engineering complex

W. A. Stevenson
Leyland Vehicles Technical Centre[†]

INTRODUCTION

The Leyland Vehicles Technical Centre consists of a test track covering approximately 40 ha and a complex test building. The Centre was designed, and it is used, for the testing of commercial vehicles and their components.

This chapter is based on the experience gained by the author, as Project Manager, during the design, construction and maintenance of the Centre which cost £22 million.

THE DESIGN BRIEF

The importance of the brief has often been stressed. It is a most valuable record of the client's requirements. It must describe the facilities required in as much detail as is reasonably possible, and it should always be up-dated. If an inadequate brief results in a building which fails to meet the requirements, then it is the client on whom the blame falls.

The brief, starting with a general description of the project, needs to be broken down into sections, which must provide all the performance information required by the consultants. Up-dated copies must be available to all key personnel who need to be aware of the contents of the document.

THE CONSTRUCTION PACKAGE

The Technical Centre was completed in a little over two years. It consists of a

† Now with Bellway Urban Renewal (Northern) Ltd.

40 ha test track, three connected buildings to house vehicle workshops, laboratories and test equipment, together with some substantial test rigs. The contract was split into three principal parts — the test track, the buildings, and the test rigs. The rig installation was controlled separately by Leyland Vehicles and was installed during the building construction, saving, as a result, a great deal of time, but creating many contractual and physical difficulties. Much pre-planning and documentation, followed by strong on-site co-ordination is required to ensure that splitting up the contract in this way can succeed.

THE CONSTRUCTION PROGRAMME

At the tender stage, the main contractor needs to be acquainted with the full effect that the client's contractors and completion dates will have on his programme. The contractor should then be required to produce a programme in sufficient detail to show how it is intended to meet the client's requirements on time. This must cover everything — temporary and permanent services, waterproofing, dust-free and humidity and temperature control dates; as well as site access, and dates for provision of site accommodation. The absence of a detailed and carefully thought out programme, will almost certainly lead to delays and extra cost.

COMMISSIONING

The design and construction team tend to accept that a single line drawn on a programme, with the word 'commissioning', is an adequate way in which to describe a most complex process. The commissioning period is often used as a buffer zone by mechanical and electrical contractors who are late with their installation. To stop this and to ensure satisfactory monitoring of the process, the operation should be split down into its many component parts. The programme should also show the work which will still be in hand after handover. In addition, the programme should indicate the training period for client's staff, marking the date from which the client is responsible for maintenance, i.e. the handover dates. Obviously the training should be complete by this date if the client is reasonably to be expected to accept responsibility for the plant. In practice, training has often only just started by the handover date. The commissioning of plant, its operation and staff training are areas which almost always lead to difficulties, generally resulting in an unsatisfactory situation for the client. A client will have more than enough problems, moving into and making use of the new building, without having to suffer the disruption caused by building service problems.

The contractors will be present to a greater or lesser extent during the first year of use, dealing with small items of outstanding work, defects, and possibly additional work. In view of this, it is not illogical to obtain, at the time the main contract tenders are required, a supplementary tender for the first year's maintenance. Placing the first year's maintenance contract with the main contractor should have benefits both for the client and the contractor. The contractor will benefit, since the contract provides work in addition to the resolving of defects during the first year. The client benefits since the first year can be treated as a running-in period for the building. As with a new car, the first service is pre-paid. Responsibility for ensuring that the services operate correctly would then rest principally with the contractor, so removing the usual arguments between client and contractor arising from non-performance of services. It also enables the client's team to spend the first year getting acquainted with the plant by working alongside the contractor's maintenance men. Hence plant training is in fact spread over the first year. Such an arrangement, if it spares the client just one expensive servicing error or prevents a plant failure resulting in the stoppage of work, will very quickly pay for itself.

ENGINEERING LABORATORY DESIGN

Ceilings
In most laboratories, regular access will be required to the many services behind a ceiling. A great deal of research needs, therefore, to be put into finding the most accessible suspended ceiling. It is certainly well worth considering fixing a sound absorbing layer to the structure soffit, and either no ceiling at all, or a simple slatted structure.

Temperature control
The brief should state clearly if rooms require separate temperature control. Problems will certainly arise if two rooms with differing heat requirements operate on one thermostat. An obvious point, but one often missed.

Fire fighting and detection
As with all systems, the simplest is the most effective. Sprinklers are the most effective for fire fighting, and can be used to protect most if not all areas. A number of flow detectors on the system gives rapid indication of the area in which sprinklers are operating.

Smoke and heat detectors should not be so sensitive as to result in numerous false alarms. A fundamental point, often missed, is that the fire protection system, in its role as a protector of the building, should not in the process, present a hazard to human life.

Floor finishes
The choice of floor finish is generally a compromise. Much research can and should be done to ensure the right compromise is achieved.

Earth leakage protection of the power supply

Earth leakage protection is being fitted in many new buildings to meet the implied requirements of safety regulations. It is often taken out soon after occupation. Many items of equipment, by their very nature, will trip the circuits. In certain circumstances a power failure can itself be a safety hazard. If required, it is better to protect the socket outlets, such that the user is aware that there is a protected circuit and that other operations will not be affected if a failure is caused.

Room access

Doors and corridors are often best made over size to allow access for future items of test equipment or unusually large test pieces.

Ground floor

In many laboratories, the ground floor is of the greatest value because of its ease of access, particularly for heavy items. In view of this, it is sensible to ensure that it is not cluttered up with plant which could go above or below the ground floor.

Noise

Much attention needs to be paid to the control of noise both inside and outside the building. Failure to resolve noise problems during construction leads often to costly alterations on completion. If sound absorbency is to be applied to the underside of a roof, it is sensible to take account of its U-value when determining the thickness of thermal insulation required.

Drawings

It is very worth while requiring the consultants to produce pictorial drawings and/or wall elevations showing the inside of each room, with the location of benches, services, tiling, windows, etc. The user will then have a clear idea of the rooms being constructed.

Drainage

This is a subject which tends to be forgotten until it forces its presence on the occupants. The designers must be made very aware of anything which may go into the drains, even by accident. The drainage design must be the subject of detailed discussions to ensure a satisfactory and workable solution to the problem.

Dust

This is always present in new buildings and arises from dusty floors, walls, other areas and conditions outside the building. If dust is likely to be a problem to the user, the designers need to be made aware of the need to exclude dust with effect from the date of handover of the building.

Security

Very expensive packages are available, but before consulting suppliers the base principles need to be resolved. Where is the secure boundary to be formed, and what is to be kept secure from whom? It may be that the best system involves few locks and is based on a reasonably secure boundary fence, with a good closed circuit television surveillance around the building and key areas within the building. Tempting, complex electronic access systems should be treated with care; they may prove an expensive liability.

First aid

Facilities for first aid should take account of all the hazards likely to be encountered. It may be that investigation will show that a surprisingly comprehensive facility is required.

Basements and floor ducts

The designers should be made aware of the conditions required in ducts and basements. If they are to be absolutely dry then this should be understood and the cost implications made clear. Account should also be taken of the requirements of safety and health regulations, which may require the provision of full basement ventilation.

Signs

There is a modern tendency to cover buildings with notices, labels, warning signs, etc. These should be clear, co-ordinated, unambiguous and fitted before handover. For example, one should be wary of marking power points 'Clean Supply'; its value is likely to be negated by office cleaners deciding that its use refers to them.

CONCLUSIONS

To achieve at a reasonable cost a facility which is satisfactory and practical in all respects, requires a great deal of informed time and work. Much depends on the choice of the right consultants, preparation of a thorough brief and the correct choice and integration of construction packages.

The complexities of most laboratories lend themselves to management contracts. Care, however, should be taken in choosing the management contractor. It is essential that the contractor has all the back-up resources necessary to carry out the task and, in fact, appreciates the extent of the role required from a management contractor. The principal features of this role are to ensure the completion of a satisfactory facility, on time and on cost. It requires the contractor to run the contract in a manner which is in the client's best interest.

36

An appraisal of a design for a flexible teaching laboratory

T. Henney
Quorn Architects Group Edinburgh

INTRODUCTION

The Charles Darwin Building, Bristol Polytechnic was designed by the Laboratories Investigation Unit (LIU) of the Department of Education and Science (DES). The LIU is a small group of architects and related professions set up to investigate and advise on ways of improving the design and use of laboratories and their associated accommodation and services. It is sponsored jointly by DES and the University Grants Committee on behalf of all government departments who are responsible for laboratory buildings.

Since its creation in 1968 the LIU has, in addition to carrying out and publishing research into user requirements, involved itself in development work, including project design, and consultancy on live projects. It has developed with users and manufacturers a system of overhead servicing and loose furniture now installed in a wide range of educational establishments in the United Kingdom and overseas.

Developments of servicing and furniture reflect the LIU's design philosophy which is based on making the laboratory responsive to changes in user needs as they evolve during the total life of the building. With this approach the building becomes a resource which can be adapted and reallocated in a variety of ways either to meet organisational changes or individual demands for particular layouts.

The Charles Darwin Building, which was commissioned by Avon County Council, offered the opportunity to test these strategies and has been fully described[1]. The LIU recognises the importance of feedback and commissioned an architect in private practice with experience in the design of laboratories to carry out an independent appraisal of the building.

The aims of the appraisal were:

to examine the briefing, design and construction process which attempted to distinguish between the basic and adaptable parts of the building;

— to appraise the planning of the various elements of accommodation and the relationship between the basic and adaptable parts of the building;
— to carry out a detailed assessment of the performance of the building fabric, finishes, environmental services, furniture and fittings, demountable partitioning and adaptable services;
— to assess the occupants' reaction to the design and operation of the building;
— to make a cost effective study of the use of the adaptable system;
— to make an assessment of the running costs of the building.

The objective of the study, carried out from September 1978 to August 1979 during the third academic year of its use, was to give a balanced appraisal of the building in use, including the occupants' reactions. This took into consideration the results of the physical and social surveys and, during the period of the study, the inputs from the design team, the users of the building and the advisory panel. The Science Department had maintained records of all the changes that had occurred since it moved into the building.

The physical survey was carried out during January and February 1979. The positions of laboratory furniture and equipment and any changes in partitions and doors were plotted and compared with the contract layout drawings. The various categories of change were recorded and quantified. All the laboratories and ancillary rooms, offices and classrooms were surveyed to assess the condition of laboratory furniture, components, materials and finishes. A social survey was carried out amongst the staff and students regularly using the building by Fieldwork Scotland, a specialist survey contractor, to obtain their reactions to the design and operation of the building. Different questionnaires were used for staff and for students. A space utilisation survey of the teaching rooms was carried out for DES by Dr Kenny, a member of the staff of that Department. During the week of the survey each identified teaching space was visited once during each hour from 0900 to 1700 and the number of students present in each hour was noted.

THE BRIEF

The basic brief for the building was to provide accommodation for the Science Department of Bristol Polytechnic but also to include a small amount of accommodation for the Mathematics Department. The Building was designed to accommodate 843 full-time equivalent students, 100 teaching staff and approximately 60 technical and administration staff.

The briefing stage provides a good two-way exchange of information between the client and the designer. The Department supplied LIU with information on projected courses, numbers of students and other data to enable calculations of areas and the number of rooms to be made. The Unit for Architectural Studies at University College London, on behalf of LIU, used their computer program to produce a schedule of the minimum number of laboratories and classrooms needed to provide sufficient working places for all students.

A small working party was appointed by the client to negotiate with LIU. It was necessary to reconcile the aspirations of the designers, who wanted to develop new concepts of flexibility, and the more traditional views of the majority of staff. The working party visited universities and polytechnics where adaptable furniture and overhead services had been used. The visits to other laboratories and the discussions with the people working in them reassured the members of the working party that adaptable laboratory furniture and overhead services were a feasible proposition.

An existing laboratory at the Ashley Down site was fitted out as a prototype with overhead services and laboratory furniture as proposed for the new building. This was used as a working laboratory by staff and students. The performance of components was assessed and feedback information provided for LIU to make modifications to the design.

The LIU systems designer prepared furniture layouts which were agreed by the staff. In the application of the LIU basic supplementary approach there was a separate contract from the main building contract for the manufacture and supply of laboratory furniture. LIU invited tenders for laboratory furniture many months after the start of the contract. The extra time for finalising the design of furniture and the layouts of laboratories was put to good use; it allowed the detailed brief to be updated as construction proceeded and reduced the likelihood of immediate changes on occupation. Cost planning and control were rigorously applied to keep the cost of the various elements within the cost target for the building.

CONTRACTUAL INFORMATION

The gross floor area of the building is $10\,613$ m^2 with a usable floor area of 7276 m^2. The tenders for the project were received in January 1973, work started on site in spring 1973 and the building completed in autumn 1976. The main contractors were George Wimpey and Co. Ltd. for a contract sum of £1 373 168. The contract sum for laboratory furniture and fittings was £183 733. Details of costs and the project team are given in Table 36.1.

BUILDING DESIGN

The site
The new site for the Polytechnic at Cold Harbour Lane was in the north-eastern outskirts of Bristol, on open ground sloping to the east. The development plan for the site as a whole had already been drawn up by the City Architects Department. A position for the Science Department had been allocated between the Mathematics and Engineering Departments with which it has a working link.

Table 36.1

Charles Darwin Building, Bristol Polytechnic: project team and details of cost

Project team	
Architects	LIU in consultation with Avon County Council
Landscape architect	Colvin & Moggeridge
Consulting engineers (services)	LIU
Consulting engineers (structural)	Clarke Nicholls & Marcel
Quantity surveyors	Gleeds (Bristol) in consultation LIU
Furniture designer	LIU in consultation with County of Avon Furniture Design Office
Main contractor	George Wimpey & Co. Ltd
Services sub-contractor	Crown House Engineering Ltd

	Cost	
	Percentage of total	Per unit gross floor area ($£/m^2$)
Substructure	6	7.10
Superstructure	39	50.11
Internal Finishes	9	11.63
Fittings and Furniture	2	2.60
Services	44	55.68
Total excluding external works		127.12
Total including external works		127.65
Laboratory furniture and fittings		18

There were certain constraints on the site, among which were that a direct covered way was to be provided linking both Mathematics and Engineering buildings with a communal building immediately to the west; the building should also contain part of the main covered circulation route running through the Polytechnic and should provide a main pedestrian entrance to the site from the east, and the floor to floor height of 3.6 m in adjacent buildings should be maintained to allow unrestricted circulation of trolleys, etc.

The building

The L-shaped plan of the building followed the general pattern of development for the site. It allowed convenient links at each end at an upper level with adjacent buildings, and also enabled a covered connection to be made with others to the south and west. The main Polytechnic circulation route is carried through the middle floor in the form of a concourse looking on to the inner court. Floor plans are shown in Figs. 36.1–36.4.

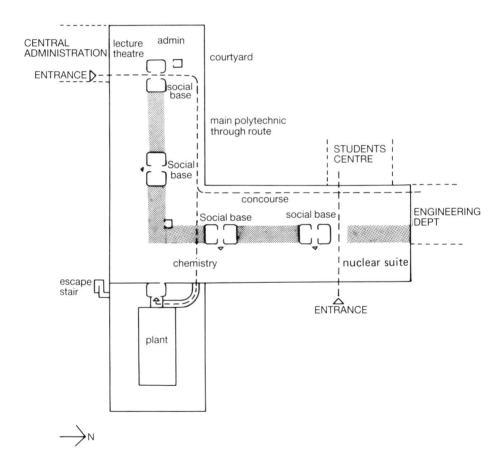

Figs. 36.1 to 36.4 – Charles Darwin Building, Bristol Polytechnic. Basic floor plans showing the central location of stairs and the core for ancillary spaces. The middle level contains the major horizontal circulation route – the concourse – and local social bases associated with each stair.

Fig. 36.1 – Ground floor – Level 1.

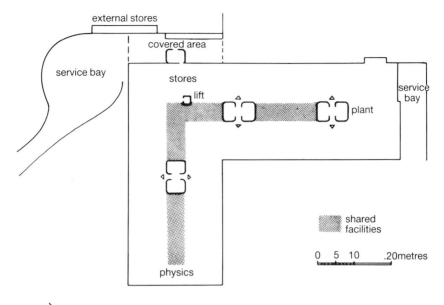

Fig. 36.2 — Lower Ground floor — Level 0.

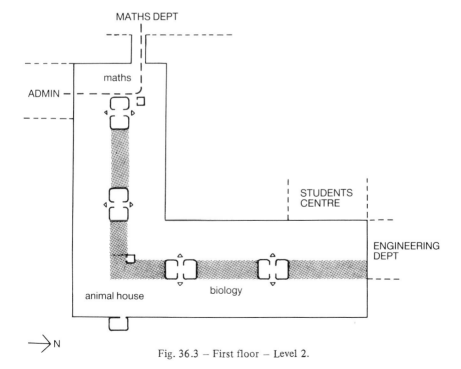

Fig. 36.3 — First floor — Level 2.

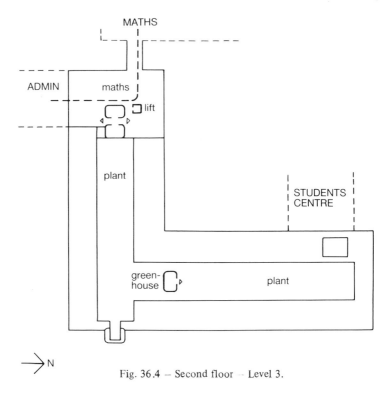

Fig. 36.4 — Second floor — Level 3.

Laboratory and general teaching accommodation was planned on three main levels — Physical Sciences on Level 0, Chemical Sciences on Level 1 and Biological Sciences on Level 2. A small amount of accommodation was also provided on Level 3, adjacent to Phase 1 Merchant Venturers Building, which is the central administration building for the Cold Harbour Lane site, for the Mathematics Department.

Shared use and change of use are allowed for by positioning supporting ancillary rooms such as special instrument rooms, preparation and storage areas in a central core zone which is directly accessible from the perimeter areas. The perimeter areas also contain staff rooms and lecture/seminar spaces related and integrated into laboratory areas. This enables joint use of lecture/seminar spaces and laboratories if desired; the location of lecture/seminar spaces adjacent to staircases also give direct access for independently time-tabled use.

Although the building extends through three floors, communication between them is readily achieved with the provision of stairs at only 25.2 m intervals in the core zone. This meets the requirements for fire escape from *en suite* planned areas. The stair enclosures contain toilet facilities and are linked horizontally at Level 1 by a major circulation route, the concourse, which provides routes through the Science Building to adjoining buildings.

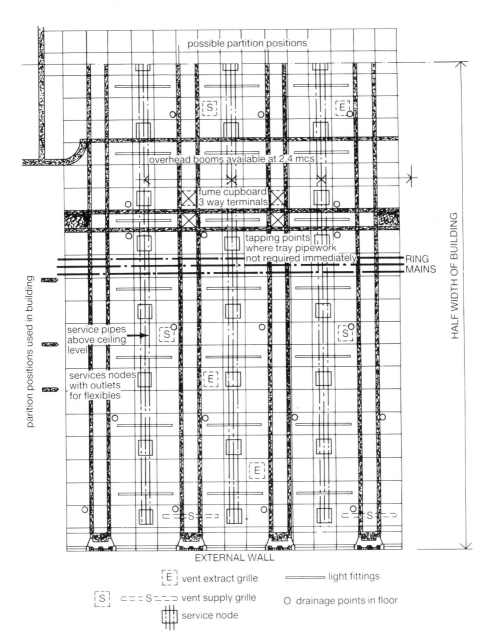

E vent extract grille   ━━━━━ light fittings

S ⊏⊏⊏S⊐⊐⊐ vent supply grille   O drainage points in floor

service node

Fig. 36.5 – Part plan – structure – services partition grid. Relationship of services distribution grids to 600 mm planning grid. Overhead services and lighting are co-ordinated to external window positions leaving a clear 600 mm wide zone behind the window mullion for two possible partition locations.

The resultant plan form, 25.2 m deep, is mechanically ventilated throughout with six to eight air changes per hour. Mechanical ventilation demands considerable space for its production and the plant for this is housed both at Level 0 next to the boiler room and at Levels 1 and 3 in roof-top glasshouses.

The interior of the building is designed on a 600 mm planning grid. Structural bays, 8.4 × 8.4 m with a 300 mm deep waffle slab of *in situ* reinforced concrete, provide a repetitive structure incorporating its own fire resistance. Main areas of each floor are sub-divided with blockwork walls some of which are carried up to the underside of the floor slab above to provide fire compartments across the building. Sub-divisions within these main areas are formed by demountable partitions which can be relocated as required without damage to floors or ceilings; floor finishes and ceilings are continuous between blockwork walls. A part plan of the structure − services partition grids is shown in Fig. 36.5.

Furniture and its associated services have the greatest potential for adaptability in the building. The mobility of furniture is linked to the provision of an overhead distribution of services incorporated in a metal tile, suspended ceiling. This provides, at each level, a regular grid of services outlets which are tapped into as required with flexible connections from furniture and equipment. The ceiling grid is matched by a grid of drainage points in the floor to which drainage from furniture and equipment is flexibly connected. The runs of services (supplies and drainage) are connected to a ring mains system routed round the building in the ceiling void at each floor. These ring mains are served by vertical risers contained within the stair toilet enclosures. Vertical distribution is similarly contained.

A range of loose laboratory furniture was designed by the LIU in consultation with the Science Department and developed in parallel with the building. A wide variety of laboratory layouts is possible, including perimeter, peninsula or island configurations. This allows users to tailor layouts to particular educational and management needs both immediately and in the future. Furniture layout is shown in Fig. 36.6. Items are not fixed to walls or partitions but hung from an aluminium picture rail which is provided in all spaces.

The servicing of laboratory areas by technicians is of paramount importance. On each floor level, preparation and washing-up areas are planned as central facilities and particular attention has been given to bulk storage, service trolleys and washing-up facilities.

Laboratory furniture

A coordinated range of laboratory furniture was used in the building. The loose furniture approach in laboratories made it easy to introduce special items of furniture when necessary. Large, floor-mounted items of equipment were located in the ancillary spaces shared by general teaching laboratories. Figures 36.7 and 36.8 show details of some of the laboratory furniture. Although adaptable furniture had been used in other laboratory buildings, adaptability covers all the furniture elements used in the Charles Darwin Building.

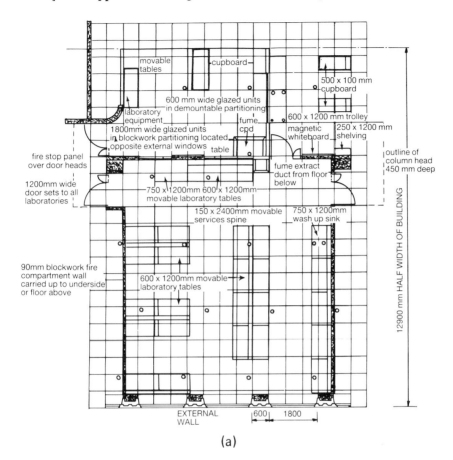

movable tables

cupboard

500 x 100 mm cupboard

600 mm wide glazed units in demountable partitioning

laboratory equipment

fume cpd

600 x 1200 mm trolley

1800mm wide glazed units in blockwork partitioning located opposite external windows

magnetic whiteboard

250 x 1200 mm shelving

table

fire stop panel over door heads

fume extract duct from floor below

outline of column head 450 mm deep

1200mm wide door sets to all laboratories

750 x 1200mm 600 x 1200mm movable laboratory tables

150 x 2400mm movable services spine

750 x 1200mm wash up sink

90mm blockwork fire compartment wall carried up to underside or floor above

600 x 1200mm movable laboratory tables

12900 mm HALF WIDTH OF BUILDING

EXTERNAL WALL

600 1800

(a)

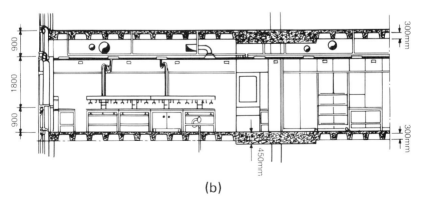

900   1800   900

300mm

300mm

450mm

(b)

Fig. 36.6 – Furniture layout (a) plan (b) section building/furniture dimensional co-ordination. Perimeter, peninsula or island configurations can be achieved with the same basic set of components based on a 600 mm planning grid.

Fig. 36.7 – Laboratory furniture in an electronics laboratory. The apparatus rack fits on top of the bench and has multiple electrical sockets (reproduced by permission of Bristol Polytechnic).

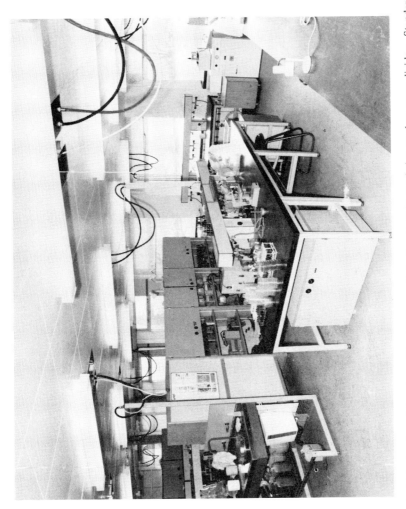

Fig. 36.8   Typical biochemistry laboratory. The 1800 mm high storage units are being used as room dividers. Standard laboratory tables, services spines and underbench storage units. Note the flexible overhead services connections to the masts on service spines (reproduced by permission of Bristol Polytechnic).

Laboratory tables
Basic 1200 × 600 mm laboratory tables, 850 mm high, some with Iroko finish, others with plastic laminate finish and a few with stainless steel tops. Some 750 mm high tables provided for sit-down work. The tables are well designed and substantially constructed and are standing up well to laboratory usage.

Services spine
Standard spine unit — 2400 × 150 × 850 mm high to worktop level; the height to top of the spine is 1300 mm. The spines are plumbed and contain all the piped services, drainage pipes and wiring. 2300 mm high service masts are located at one end and are connected with flexible connections to overhead service nodes. A flexible PTFE tube is attached to the glass bottle trap and connects to a drainage outlet in the floor. The spine is of compact design and the services well integrated into its tight dimensional constraints. The services spine is one of the key elements in the adaptable furniture system.

Bench stools
The tractor seats are severely criticised by the users who consider them to be noisy and unsafe. Some of the stools, non-stackable, have suffered from metal fatigue.

Underbench storage cupboards
Standard units from a proprietary range. The cupboard is easy to manoeuvre and used extensively as a free standing unit to support equipment.

Free-standing storage units
Four models, 1000 × 500 × 1800 mm high, from a proprietary range of laboratory units have been used. These are single-sided with deep shelves, double-sided with shelves, full height cupboards, or cupboards in the upper half with shelves below. These units are well designed and constructed. Each unit is a discrete item designed to be free standing. The storage units are used successfully throughout the building.

Stainless steel sink units
Custom designed with a large, stainless steel bowl, overall size 1200 × 750 × 950 mm high, with a service mast to take water and electrical supplies, and having flexible connections to overhead ceiling nodes. Each unit has a 1 kW, 10 litre capacity electric, water heater. The sinks designated for washing-up activities are plumbed in with a hot water supply. This is a reasonably successful laboratory furniture unit.

Top rail support
Wall shelving, cupboards, pinboards, chalkboards, etc. are supported on a continuous horizontal top rail which has the same function as a picture rail. Positioned at door height, 2100 mm, this has proved to be a successful support system. Its range and flexibility have been fully exploited. The laboratory users have responded by extending its usage to meet new requirements.

Fume cupboards
Custom designed mobile cupboards. Glass on all sides of the fume cupboards make them fully transparent for demonstration work. The full range of flexible connections for direct linkage to the fume extract, services supply and drainage system is incorporated.

The fume cupboards have been designed to operate as part of the air extract system and have an extract performance of 0.5 m/s when the sash is opened to 500 mm above the cill.

RESULTS OF THE USER SURVEY OF LABORATORY FURNITURE

Fifty-four per cent of the staff and sixty per cent of students gave a satisfactory rating to the use of adaptable furniture. The staff were very satisfied with the laboratory tables, but bench stools were considered to be poor by both staff and students. Sinks were criticised by users on a number of points of detail. A major criticism was the complete lack of drawers in the underbench storage units and there was minor criticism of the design of the fume cupboards.

INTERIOR DESIGN

The LIU made a great effort with the interior design of the building to provide variety and visual impact. Colour was used positively. The focal nature of the stair/toilet enclosures is emphasised by giving each a strong identifying colour which is echoed with variations on blockwork walls running from the core area to external walls. The permanent features of the building – the concourse, the stair/toilet enclosures and the main lecture theatre have been distinguished from the general areas by detailed design and furnishings. The general condition of the internal areas was found to be very good.

HEATING AND VENTILATION

The walls were insulated to 0.36 and roof to 0.41 $W/m^2/°C$. The windows are single glazed and amount to 38 per cent of the external wall viewed from the inside. The main building is served with three separate heating and ventilating units with separate units for the Nuclear Suite, and the Animal Suite. The units are all located in a glasshouse plant room. Mechanical ventilation provides six to eight air changes per hour. There is no recirculation of extracted air all of which is discharged to atmosphere. Fresh air is either heated or cooled in the main air-handling plant. To reduce duct sizes, air is distributed at high velocity and is converted to low velocity in terminal units linked to attenuators. The final distribution to ceiling diffusers is through flexible ducts. The ceiling space is used as an extract plenum. The air extracts from the fume cupboards have been designed as an integral part of the air-handling system.

Temperature control is considered by the majority of the occupants to be the most critical problem. There are twelve zones from each plant and since each zone covers many types of accommodation, e.g. offices, classrooms and laboratories, problems are created. If the heat sensor for a zone is located in a room with heat generating equipment it is difficult to achieve equitable ambient temperatures in other rooms. Some offices were particularly cold in winter and thermostatically controlled heaters have been installed to overcome this problem. In some of the internal core rooms temperatures of 25°C have been recorded. There appears to be mis-matching between the concept of total planning flexi-bility and the ability of the air-handling system to respond to these changes. There is sufficient heating, cooling and air movement capacity in the system, but it needs to be broken down into a greater number of controlled zones. The strict cost constraints were a major factor in the design of the system and, if more money had been available for more controlled zones and instrumentation, the problem could have been alleviated.

In the user survey 80 per cent of the staff and 55 per cent of the students were dissatisfied with temperature control; 50 per cent of the staff and 55 per cent of the students were dissatisfied with the ventilation. All other environmental aspects were considered satisfactory.

ASSESSMENT OF THE PLANNING OF THE BUILDING

Physical survey

The physical survey revealed that gangway widths between benches are in some places inadequate. These vary between 1050 and 1800 mm, the average being 1,400 mm. In some cases the gangway had been reduced to 800 mm. The survey showed that the considerable amount of equipment in the laboratories is putting pressure on the available floor space and gangways. Each laboratory is well-equipped with deep freezers, refrigerators, ovens and centrifuges. Bench-mounted equipment, water baths, and centrifuges take up bench space. The designers had envisaged that laboratory equipment housed in core ancillary rooms would be partly shared and moved into the laboratory when required. The laboratory area had therefore been reduced and the area of core rooms increased. Some teaching laboratories have been planned *en suite* and there are no corridor connections to adjoining rooms.

From the physical survey it was evident that no additional area was allowed for inter-room circulation. The survey showed that storage normally located in the bench zone had been deployed to free-standing storage units. These units take up floor space when grouped together and require circulation space to form gangways.

There is a problem in fitting all the furniture, equipment and storage units into the laboratories. This has resulted, in some cases, in gangways which are too narrow and raises the problem of safety of movement within laboratories. It is recommended that extra floor space of 0.6 m^2 per student place is required to accommodate storage, free-standing equipment and allowance for inter-room circulation.

Local and general facilities

Preparation and service rooms are generally located in the core of the building and provide back-up for the general laboratories. These service rooms include a Nuclear Suite, Animal Suite, tissue culture and instrument rooms. These rooms are used for garaging and servicing mobile equipment ready to be moved into the adjoining laboratories. Special equipment is extensively used for teaching and project work in instrumental techniques for the mainly para-medical courses provided by the Science Department. There is a good functional relationship between the ancillary accommodation and the teaching and research/project areas.

Storage workshops are located on Level 0 and are connected by a goods lift to local services facilities on each floor. An allowance of 3.5 m^2 per student place was made for preparation rooms, special teaching areas, workshops and central stores.

Health and safety considerations

Health and safety requirements were well considered in the design of the laboratories. Great care and attention has been given to the design of fire compartments and escape routes. Precautionary and instruction notices, warning symbols, first-aid boxes, eye-wash containers, fire-fighting appliances, and fire-alarm boxes are well displayed and distributed throughout the building. Special safety and health requirements apply to the Nuclear and Animal Suites. The Department has a Safety Committee that monitors safety procedures.

Changes in use

Before the building was completed, adjoining classrooms were linked together to form one large classroom. The Metallurgy Section, which is associated with the Engineering Department, decided to remain on the original site at Ashley Down. The laboratory accommodation allocated to metallurgy was taken over as laboratories for the degree course in Environmental Health. The metallurgy preparation room was re-used to accommodate nuclear magnetic reasonance equipment. A general physics laboratory was converted into a teaching laboratory for serology, and an internal classroom into a tissue culture room.

During the physical survey, the position of furniture and equipment was noted, plotted and any changes in doors and partitions from the original layouts also recorded. The following were recorded in the 83 rooms surveyed: 12 changes in partitions, 10 changes of use, and 60 major and 21 minor changes in furniture. In two rooms there were no changes.

The staff in the Department have made good use of the in-built flexibility provided by the system of demountable partitions, overhead services and relocatable furniture. The design philosophy recognised the need to provide for the indeterminate future requirements of the Science Department. An example of change of use is illustrated in Figs. 36.9 and 36.10.

The cost of changes

During the period July 1976 to September 1978, 54 changes, an average of 24 changes per year, varying widely in degree of complexity were carried out and recorded. Four of the most significant changes were selected and costed. Realistic estimates were made of the expenditure on each, as if the original installations had been traditionally fixed and tailor-made requiring the employment of an outside building contractor. These were compared with actual costs incurred, i.e. technicians' wages plus an allowance for establishment charges. The result is shown in Table 36.2.

Table 36.2

Comparison of theoretical and actual cost of changes in laboratory layout

| | Cost (£) | |
	Theoretical	Actual
Total for 4 changes	7500	214
Average	1875	54
Average difference	£1821	

The average difference in cost in these examples is very marked but there is no suggestion that the magnitude of change that they entail is typical. Some of the changes recorded, for example the relocating of items which, in the traditional situation, would also be freely moveable, would show no difference in cost.

Taking the changes as a whole it would be reasonable to expect that the average differential cost might be £800. At 24 changes per year this could result in a total excess annual expenditure of £19,200. However the cost of making

Fig. 36.9 — Optics laboratory before change. There was insufficient bench space, resulting in equipment being constantly moved as class requirements changed (reproduced by permission of Bristol Polytechnic).

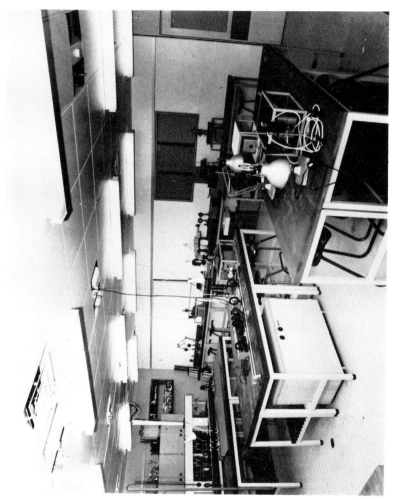

Fig. 36.10 — Optics laboratory after change. As water and gas was not required the services spines were removed, and benches re-arranged to form demonstration benching (reproduced by permission of Bristol Polytechnic).

changes is the sum of the cost of the alteration works and the cost of the disruption to normal activities caused by these works. The latter cost is fairly easily calculated for a production laboratory, but in the context of education any assessment must be speculative.

The Department's autonomous control of the adaptable elements of the building, and the considerable cost saving in carrying out changes compared with the more traditional laboratories are the two most significant features in the building. The changes can be made quickly and at comparatively low cost. The design and operation of this building should be of considerable interest to anyone involved in the design and operation of educational and research establishments.

SUCCESSFUL FEATURES

The approach to laboratory design
The design of the building is a significant milestone in the development of new concepts in laboratory design. The overall view of staff and students is that the building is a great success and the design approach has been vindicated by a close examination of the building in use.

Change and growth
Change and growth has been provided for by the systems of adaptable furniture and services and removable partitions. The major breakthrough is the development of an integrated system of overhead services in conjunction with a network of floor drainage points. The services are disassociated from partitions which can, therefore, be relocated easily. The standard overhead services network covers the complete internal floor area of the building and provides the potential for adapting more spaces into serviced laboratories.

Laboratory services
The distribution of ceiling service nodes and floor drainage points have met all user demands to date. The process services to the laboratories have been found to be satisfactory.

Constructional process
The basic/supplementary approach has been successfully applied in the building. The fabrication of laboratory furniture off site rather than on site has had considerable cost benefits, has simplified the constructional process and did reduce the fitting-out time. The integration of suspended ceilings and overhead services demands a high level of co-ordination both in design and construction; this has been achieved but not without difficulty.

Supplementary elements
The rational approach adopted in the design of a standard range of laboratory furniture has helped the Science Department with the initial fitting out of the building and with subsequent changes. The standardised range of furniture is used by the biology, chemistry and physics disciplines and by all teaching and research activities in the building. The adaptable furniture and service connections are standing up well to laboratory use.

Interior design
The judicious use of colour and materials has created a pleasant internal environment. The interior of the building is in remarkably good condition, is standing up to wear and tear and there is no evidence of vandalism.

Research/project areas
The adaptability of the system is being fully utilised by research workers.

Local services facilities
These facilities which provide back-up services to the laboratories are well-positioned in the core of the building. The space in some rooms is rather congested but the overall space allocation for these facilities is about right.

General services facilities
The main stores, workshops and central preparation areas have been well planned and operate efficiently. The technical staff are satisfied with the facilities.

Social areas
These are positioned off the main concourse, are popular with students and help to make the building a pleasant place to work and study in.

The building in use
In the social survey most aspects were rated as satisfactory by a large proportion of the respondents. Of the seven aspects covered by the students, five were rated satisfactory by 50 per cent or more. Of the fifteen aspects covered by the staff, thirteen were rated satisfactory by 50 per cent or more.

Changes in use
The first three years are too short a time in the life span of the building to assess if its adaptability will be extensively used. During this period the micro-flexibility – the movement of furniture and equipment – has been well used. It is difficult to predict to what extent the macro-flexibility will be utilised. Continuing feedback reports from the Polytechnic indicate that further changes are taking place and substantial changes are projected.

Cost of changes

The ability to make rapid changes to the layout of the laboratories and offices at a low cost compared to conventional accommodation is a considerable advantage arising from the flexible design strategy.

FEATURES WHICH COULD BE IMPROVED IN SUBSEQUENT DESIGNS

Space for services

The 600 mm space above the ceiling for services is too restricted, and the designers had considerable problems in fitting all the services within this space. It is recommended that the space provided for a similar range of services should be 900 to 1000 mm.

Heating and ventilation

The occupants' major complaint was the difficulty in achieving equitable temperature control in the various rooms in the building. The heating and ventilating system for this building was designed on a broad-brush basis and does not have the in-built flexibility to match the adaptability of the LIU system.

More supply zones and control instruments are necessary and consideration should be given to heat recovery measures as the discharge of all the extracted air is not energy efficient.

Location of fume cupboards

The vertical riser ducts for the fume cupboards are located in a zone adjoining the core rooms. The positions of fume cupboards conflict with the main circulation and escape routes from the laboratories.

Laboratory space

The space in the laboratories cannot easily accommodate the amount of furniture and equipment in use. This is attributed to three points.

(1) The amount of free-standing and bench-mounted equipment required in laboratories was not envisaged at design stage.
(2) There is a lack of bench-related storage and an inadequate allocation of space for free-standing storage units.
(3) Where a teaching laboratory was planned *en suite* with adjoining rooms, no extra space was allocated to the laboratory for inter-room circulation.

These points have led to the width of gangways between benches being too narrow for safe movement of personnel. Space standards, for this type of laboratory should be reconsidered, and an increase to 0.6 m^2 per student place is recommended.

Offices

The staff are not satisfied generally with the standard of office furniture. Inadequate temperature control and the poor sound reduction property of the partitions between the rooms also cause dissatisfaction.

Current utilisation

The original[1] concept assumed a population of 843 full-time equivalent students, and predicted a 50 per cent utilisation of laboratories and 75 per cent utilisation of classrooms over a 32-hour week. The actual student population was 611 full-time equivalent students for the academic year 1978–79. The utilisation survey recorded 11 per cent utilisation of laboratories and 32 per cent utilisation of classrooms over a 32-hour week.

OPERATION OF THE BUILDING

It was the designer's intention that the Science Department had the autonomous control of their building in implementing changes. It was disappointing that LIU was unable to provide an operational manual as had been intended at the time of occupation, although an operational manual has now been prepared. There is a limited stockpile of spare components and furniture, but not enough for this type of adaptable building. The lack of supporting logistics could be a problem in the implementation of future changes. Initial problems of storing surplus furniture and fittings have now been resolved by using a large internal lecture room for this purpose.

REFERENCE

[1] *The Charles Darwin Building, Bristol Polytechnic,* Laboratories Investigation Unit, Bulletin No. 9, Department of Education and Science, HMSO, London, 1977.

Index

Since everything in this book relates in some way to laboratories, very little has been indexed under this heading, except where it seemed perverse not to. Such topics as safety, cost, and lighting are therefore under themselves.